FUNDAMENTALS OF EDDY CURRENT TESTING

DONALD J. HAGEMAIER

THE AMERICAN SOCIETY FOR NONDESTRUCTIVE TESTING, INC.

Published by The American Society for Nondestructive Testing, Inc., P.O. Box 28518, Columbus, OH 43228.

ISBN 0-931403-90-1

Printed in the United States of America.

Foreword*

"The introduction by [Dr.] Förster of sophisticated, stable quantitative test equipment and of practical methods for analysis of quantitative test signals on the complex impedance plane were by far the most important factors contributing to the rapid development and acceptance of electromagnetic induction and eddy current tests during 1950-65 in the United States. Förster is rightly identified as the father of modern eddy current testing.

"By 1950, he had developed a precise theory for many basic types of eddy current tests, including both absolute and differential or comparator test systems and probe or fork coil systems used with thin sheets and extended surfaces. Each test was confirmed by precise solution of Maxwell's differential equations for the various boundary conditions involved with coils and test objectives, at least for symmetrical cases such as round bars, tubes, and flat sheets where such mathematical integrations were feasible.

"By extensive publications, Förster made the results of this research available to the world of technical personnel. His contribution of almost the entire theory and technology of electromagnetic induction and eddy current test techniques to the first edition of ASNT's *Nondestructive Testing Handbook* provided the means for educating thousands of other NDT personnel in the theory, methods, equipment, and interpretation of eddy current tests. This integrated presentation was then used throughout the world to update eddy current test technology.

"In the United States, numerous facilities began research to test these new concepts and instrumentation, including significant efforts at Oak Ridge, Hanford, and other facilities. The creative work of Hugo Libby at Hanford, WA, during the past 30 years, and that of Robert Oliver, Robert McClung, Caius V. Dodd, J. A. Deeds, and others at Oak Ridge, TN, may well have been sponsored in response to the original work done by Förster."

From this research work came the eddy current instrumentation that made the technology available to various industries. Developers of these instruments are credited to Glenn L. McClurg (Magnaflux Corp.), Charles Fredrick (Nortec Corp.), Clyde Denton (Zetec Corp.), Richard Hochschild (Budd Co.), and Donald Hocking (Hocking Electronics).

---

*Excerpted from R. C. McMaster's article, "The Origins of Electromagnetic Testing," published in ASNT's journal, *Materials Evaluation*, 1985, Vol. 43. No. 8, pages 946-956.

Preface

This book was prepared for The American Society for Nondestructive Testing, Inc. (ASNT) for instruction in the subject of eddy current testing (ET).

Eddy current testing is one of the most widely applicable of the nondestructive testing methods. It depends on measuring the changes in the impedance of a coil due to changes in the flow of eddy currents induced in a conductor. Any material change that affects the flow of the induced eddy currents sufficiently can be detected. Because so many things affect this flow, eddy current testing can be applied to a wide variety of test situations. Principal areas of application are flaw detection, material-thickness measurements, alloy sorting, and the monitoring of metallurgical conditions such as hardness and heat treatment. Besides its versatility, the major advantage of eddy current testing is the speed at which tests can be performed. It can be used also to inspect both nonmagnetic and ferromagnetic materials. The principal disadvantage of the method is its limited depth of penetration into the material or part being inspected.

James Clerk Maxwell conceived and published the comprehensive group of relations for the electromagnetic field known as "Maxwell's Equations," which mathematically represent almost the entire present knowledge of this subject. These equations are beyond the scope of this text. Those interested in more advanced theory are directed to the ASNT *Nondestructive Testing Handbook: Electromagnetic Testing,* Vol. 4, 2nd ed. (1986). Herein, all equations are confined to algebra and trigonometry in line with a study of fundamentals.

Acknowledgments

Material for this book was gleaned from notes on basic electrical theory and from material contained in the ASNT *Nondestructive Testing Handbook: Electromagnetic Testing,* Vol. 4, 2nd ed. (1986); the ASM/ASNT Home Study Course (Fundamentals of Nondestructive Testing), and the Atomic Energy of Canada Limited Eddy Current Testing.

The author wishes to acknowledge those authors whose work contributed to the preparation of this text:

Hugo Libby
Dr. Fredrick Förster
Dr. Robert McMaster
Michael King
Val Cecco
James Pellicer
Henry Ridder

Donald J. Hagemaier is unit chief of nondestructive evaluation (NDE) in materials and process engineering at Douglas Aircraft Co., Long Beach, CA. An active member of ASNT, he has written and lectured extensively on aerospace nondestructive testing (NDT) for more than thirty years. The recipient of the 1972 ASNT Achievement Award, he is a registered professional quality engineer with the State of California and is an ASNT Fellow in addition to being Level III certified by the Society in six methods. He has an A.S. degree in physics from Pierce College and he majored in aeronautical engineering at Northrop Aeronautical Institute.

Summary of SI Prefixes

This simplified system (SI—Systeme International d'Unités) has already been adopted with slight modifications in the legislation of most countries that use the metric system, and it is the system which is being introduced in the U.K.

These units of measurement will be the international language of mathematics, science, and technology and will eliminate the need to convert from one country's units to those of another. All other units in the system can be derived from the basic units.

The system allows units to be devised for quantities arising as multiples and submultiples of the original unit. The order of magnitude and symbol associated with these prefixes are given in the following table.

Prefix Table

Multiple	Prefix	Symbol
10^{12}	tera-	T
10^{9}	giga-	G
10^{6}	mega-	M
10^{3}	kilo-	k
10^{2}	hecto-	h
10^{1}	deca-	da

Submultiple	Prefix	Symbol
10^{-1}	deci-	d
10^{-2}	centi-	c
10^{-3}	milli-	m
10^{-6}	micro-	μ
10^{-9}	nano-	n
10^{-12}	pico-	p
10^{-15}	femto-	f
10^{-18}	atto-	a

Examples:	
megahertz	MHz
microsecond	μs
kilovolt	kV
millimeter	mm
nanometer	nm
picofarad	pF

Table of Symbols and Units

Symbol	Description	Unit
q	quantity of stored charge	coulomb
S	conductance	mho/m
σ	conductivity (sigma)	% IACS
I	current	ampere
E	potential difference between two unlike charges	volt
R	resistance; opposition to current flow	ohm (Ω)
P	power (amperes times volts)	watt
π	pi	3.1416
f	frequency (cycles/second)	hertz
T	amount of time for one cycle (i.e., period)	second
λ	wavelength (lambda); velocity/frequency	meter
Ψ	flux linkage with magnetizing coil	weber-turn
C	velocity of electromagnetic wave	m/s
ϕ	magnetic flux (encircled by primary turns)	weber
B	flux density	gauss
μ	magnetic permeability (mu); B/H	
H	magnetizing force	oersted
L	inductance; opposition to change in current	henry
k	coefficient of magnetic coupling	
X_L	inductive reactance in AC circuit	ohm
Ω	angular frequency (omega); 2π times frequency	radian
Z	impedance	ohm
N	number of series windings in magnetizing coil	
δ	eddy current standard depth of penetration (delta)	cm
θ	phase-angle lag (theta); x/δ	radians/s
ρ	electrical resistivity (rho)	
Δ	change in any parameter (uppercase delta)	
η	fill factor (eta) between coil and part	
Z	impedance vector or phasor	ohm
f_g	limit frequency	Hz
f/f_g	frequency ratio	
W	tube-wall thickness	cm
d_i	inside diameter of tube	cm
D	coil diameter	cm
d_a	outside diameter of tube or bar	cm
$\omega L/\omega L_0$	normalized inductive reactance	ohm
$\omega R/\omega L_0$	normalized resistance	ohm
E_{imag}/E_0	normalized inductive reactance	ohm
E_{real}/E_0	normalized resistance	ohm

Math Symbols and Terms

In mathematics, Greek letters are often used, particularly for angles. Some of the more commonly used ones, but not the whole alphabet, are the following:

alpha	α	mu	μ
beta	β	nu	ν
gamma	γ	pi	π
delta	Δ δ	rho	ρ
epsilon	ϵ	sigma	Σ σ
eta	η	phi	ϕ
theta	θ	psi	Ψ
lambda	λ	omega	Ω ω

Contents

Chapter 4 - Electromagnetism

Chapter 5 - Induction by Alternating Current

Chapter 6 - Inductive Reactance and Impedance

Chapter 7 - Eddy Current Test Principles

Chapter 8 - Effects of Magnetic Coupling with Test Materials

Chapter 9 - Coils, Instruments, and Standards

Chapter 10 - Impedance-Plane Response

FUNDAMENTALS OF EDDY CURRENT TESTING

Chapter 1
Electrical Theory

Electrons and Protons in the Atom

Although there are any number of possible methods by which electrons and protons might be grouped, they assemble themselves in specific combinations that result in a stable arrangement. Each stable combination of electrons and protons makes one particular type of atom. For example, Fig. 1-1 illustrates the electron and proton structure of one atom of the gas hydrogen. This atom consists of a central mass called the nucleus and one electron outside the nucleus. The proton in the nucleus makes it the massive stable part of the atom because a proton is 1,800 times heavier than an electron.

In Fig. 1-1, the one electron in the hydrogen atom is shown in an orbital ring around the nucleus. To account for the electrical stability of the atom, consider the electron as spinning around the nucleus as planets revolve around the sun. The electrical force attracting the electrons in toward the nucleus is balanced by the mechanical force outward on the rotating electron, maintaining the electron in its orbit around the nucleus.

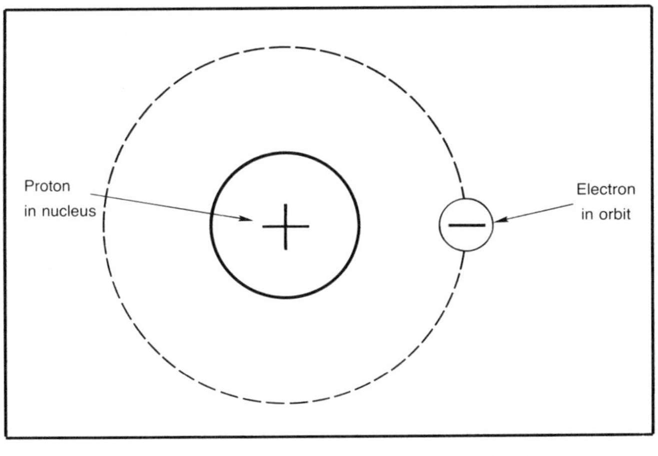

Figure 1-1. Electron and proton in the hydrogen atom.

In an atom that has more electrons and protons than hydrogen, all the protons are in the nucleus, and all the electrons are in one or more outer rings, or shells. For example, the carbon atom illustrated in Fig. 1-2(a) has six protons in the nucleus and six electrons in two outer rings. The total number of electrons in the outer rings equals the number of protons in the nucleus.

The distribution of electrons in the orbital rings determines the atom's electrical stability. Especially important is the number of electrons in the ring

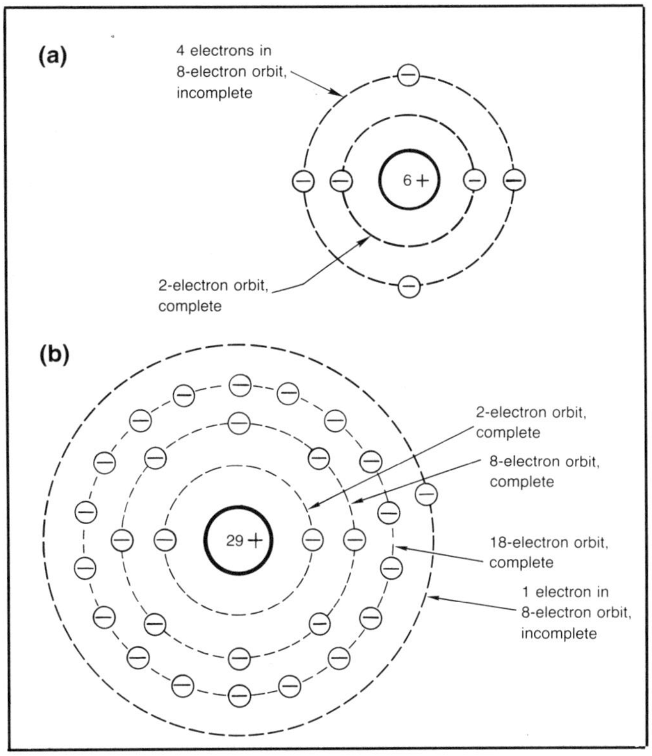

Figure 1-2. Atoms with more than one orbital ring. (a) Carbon atom with six protons in nucleus and six orbital electrons. (b) Copper atom with 29 protons in nucleus and 29 orbital electrons.

farthest from the nucleus; instability of the electrons in the outer ring makes the atom a good conductor of electricity.

The outermost ring of any element requires a maximum of eight electrons for stability, except when there is only one ring, which has a maximum of two electrons. In the carbon atom in Fig. 1-2(a), with six electrons, there are just two electrons in the first ring because two is its maximum number. The remaining four electrons are in the second ring, which can have a maximum of eight electrons. The copper atom in Fig. 1-2(b) has only one electron in the outermost ring, which can include eight electrons. Therefore, the outside ring in the copper atom is less stable than the outside ring of the carbon atom.

Copper is a much better conductor of electricity than carbon because the copper atom has a more unstable outer ring. The one orbital electron in each copper atom can move between atoms easily as compared with carbon because the carbon atom approaches the stable condition of a full outer ring. Also, a copper atom has more rings than a carbon atom. The outermost ring of a copper atom is thus farther from the nucleus. Therefore, the one electron in the outer ring of the copper atom is more independent of the stabilizing influence of the nucleus. As a result, when there are many atoms close together in a copper wire, the outermost orbital electrons can migrate easily from one atom to another at random. Such electrons are often called "free electrons." This freedom accounts for the ability of copper to conduct electricity very easily. It is the movement of free electrons that provides electrical current in a metal conductor.

Historically, the electron shells have come to be designated by the letters K, L, M, N, O, P, and Q, starting with the one nearest the nucleus. Fig. 1-3 shows the shells and the maximum number of electrons that may be present in each. Any one of the shells shown in Fig. 1-3 may be the outermost, depending upon the particular element under consideration. The K shell has a maximum capacity of two electrons; each of those beyond the K shell has a maximum capacity of eight electrons when it is the *outermost* shell. The inert elements, such as neon, argon, krypton, etc., have eight electrons in their outermost shells. Inner shells may acquire electrons up to the maximum capacity, as shown in Fig. 1-3.

Conductors and Insulators

When electrons can move easily from atom to atom in a material, the material is a *conductor*. In general, all the metals are good conductors, with silver the best and copper second, because their atomic structures allow easy movement of the outermost orbital electrons. A material with atoms in which the electrons tend to stay in their own orbits is an *insulator* because it cannot conduct electricity very easily. However, the insulators are able to hold or store electricity better than the conductors. An insulating material, such as glass, plastic, rubber, paper, air, or mica, is called a *dielectric*, meaning it can store an electric charge. Carbon can be considered a *semiconductor*, conducting less than the metal conductors but more than the insulators.

Table 1-1 lists some examples of elements. These are just a few out of the total of approximately 100. Notice how the elements are grouped.

Table 1-1
Examples of the Chemical Elements

Metal conductors (in order of conductance)	Silver Copper Gold Aluminum Tungsten Zinc Nickel Iron Lead
Semiconductors of electricity	Carbon Silicon Germanium Selenium
Gases (active)	Hydrogen Nitrogen Oxygen
Inert gases	Helium Neon Argon

The metals listed are all good conductors of electricity because each has an atomic structure with an unstable outer ring that allows many free electrons. The semiconductors in the next group have fewer electrons than the metals. Among the gases, those that are active chemically and electrically have an atomic structure with an incomplete outer ring. The inert gases have a complete outer ring, which makes them chemically inactive.

Negative and Positive Charges

If you rub a hard rubber pen or comb on a sheet of paper, the rubber will attract a corner of the paper if the paper is free to move easily. The paper and rubber thus give evidence of a static electric charge. The work of rubbing resulted in separating electrons and protons to produce a charge of excess electrons on the surface of the rubber and a charge of excess protons on the paper.

Because paper and rubber are dielectric materials, they hold their extra electrons or protons. As a result, the paper and rubber are no longer neutral; each has an electric charge.

The resultant electric charges provide the force of attraction between the rubber and the paper. This mechanical force of attraction or repulsion between charges is the fundamental method by which electricity makes itself evident. The charge is *static electricity* because the electrons or protons do not flow in a current. There are many examples of static electricity produced by the mechanical work of rubbing against friction. When you walk across a wool rug, your body becomes charged with an excess of electrons. Similarly, silk, fur, glass, and amber can be rubbed to produce a static charge.

The Coulomb: Unit of Charge

A charge of many billions of electrons or protons is necessary for practical applications of electricity. Therefore, it is convenient to define a practical unit called the *coulomb,* equal to the charge of 6.28×10^{18} electrons or protons stored in a dielectric. The symbol for electric charge is q, for quantity. For instance, a charge of 6.28×10^{18} is stated $q = 1$.

Negative and Positive Polarities

Historically, the *negative* polarity has been assigned to the static charge produced on rubber, amber, and resinous materials in general. *Positive* polarity refers to the static charge produced on glass and other vitreous materials. On this basis, the electrons in all atoms are basic particles of negative charge because their polarity is the same as the charge on rubber. Protons have a positive charge because the polarity is the same as the charge on glass.

Charges of Opposite Polarity Attract

If two small charged bodies of light weight are mounted so that they are free to move easily and then placed close to each other, one can be attracted to the other when the two charges have opposite polarity; see Fig. 1.4(a). In terms of electrons and protons, the bodies tend to be attracted to each other by the force of attraction between opposite charges. Furthermore, the weight of the electron is only about 1/1,800 the weight of a proton. As a result, the force of attraction between electrons tends to make electrons move to protons.

Charges of the Same Polarity Repel

In Fig. 1-4(b) and (c), it is shown that when the two bodies have an equal amount of charge with the same polarity, they repel each other. In Fig. 1-4(b), the two positive charges repel, and two negative charges of the same value repel each other in Fig. 1-4(c).

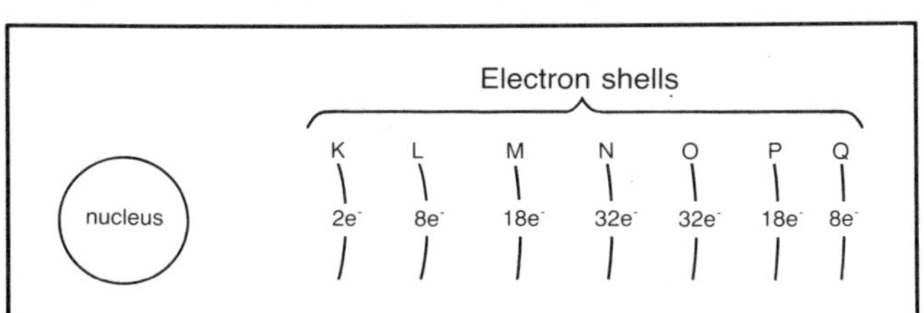

Figure 1-3. Electron shells and the maximum number of electrons that may be present in each.

Potential Difference

Potential refers to the possibility of doing work. Any charge has the potential to do the work of moving another charge, either by attraction or repulsion. This ability of a charge to do work is its potential.

A charge is the result of work done in separating electrons and protons. There is stress and strain associated with opposite charges, caused by the separation of electrons and protons, because normally those particles would be balancing each other to produce a neutral condition.

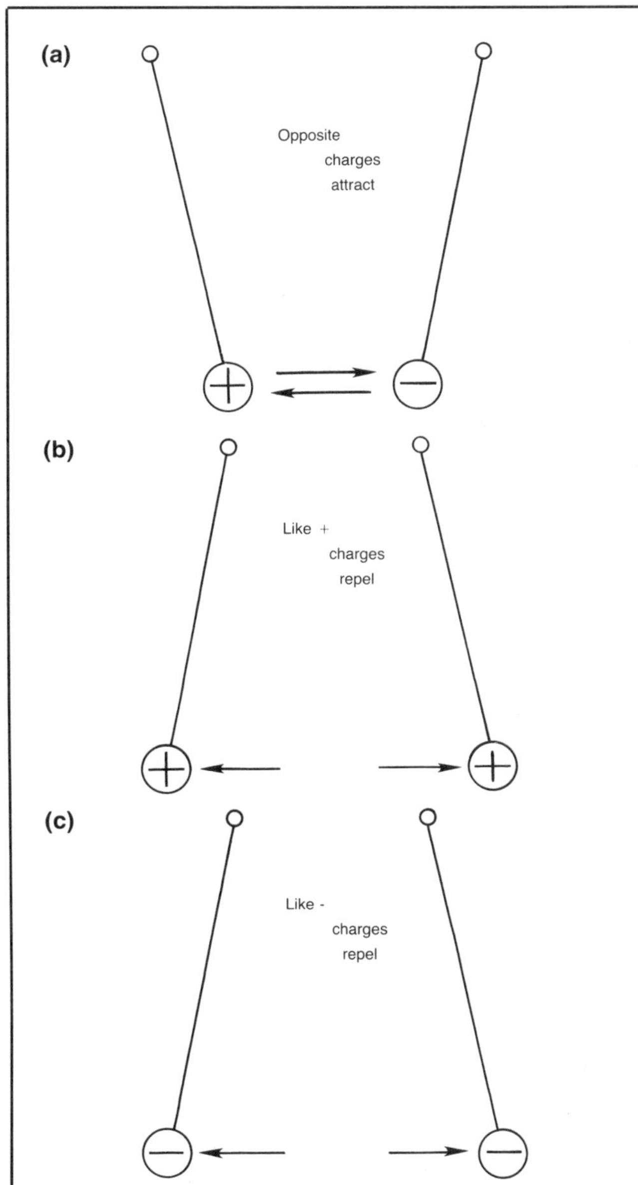

Figure 1-4. Force between charges. (a) Opposite charges attract. (b) Like positive charges repel. (c) Like negative charges repel.

Potential between Different Charges

When one charge is different from the other, there must be a difference of potential between them.

In Fig. 1-5(a), the 1 coulomb of negative charge at the left can do the work of repelling three electrons. The same amount of positive charge at the right can attract three electrons. As a result, six electrons are moved toward the positive charge. The difference of potential here is equivalent to the work of a 2-coulomb charge.

In Fig. 1-5(b), the two charges are equal. Each charge has the potential to move electrons but the difference in potential is zero. Electrons attracted toward one charge are also attracted in the opposite direction to the other charge. As a result, no electrons move to either charge because there is no potential difference between equal charges of the same polarity. To do the work of moving electrons in a specific direction between two charges, the charges must supply a potential difference.

The Volt

The practical unit of potential difference is the *volt*, which is a measure of the amount of work required for moving one coulomb of charge. When 1.03 N-m (0.7636 ft-lb) of work is necessary to move 6.28×10^{18} electrons between two points, each having its own charge, the potential difference between these two points is 1 volt. The symbol for potential difference is E, a symbol for electromotive force, indicating the ability to do the work of forcing electrons to move. Because the volt unit is used so often, potential difference is often called *voltage*. However, voltage is fundamentally the difference in potential between two points. The term emf as an abbreviation of electromotive force is sometimes used for a voltage source.

Charge in Motion Is Current

When the potential difference between two charges forces a third charge to move, the charge in motion is an *electrical current*. To produce current, therefore, the charge must be moved by a potential difference. In solid materials, such as a copper wire, the free electrons are charges that can be forced to move with relative ease by a potential difference because they require relatively little work to be moved. As illustrated in Fig. 1-6, if a potential difference is connected across two ends of a copper wire, the applied voltage forces the free electrons to move. This current is a drift of electrons from the point of negative charge at one end, moving through the wire, and returning to the positive charge at the other end.

To illustrate the drift of free electrons through the wire shown in Fig. 1-6, each electron in the middle row is numbered, corresponding to a copper atom to which the free electron belongs. The electron at the left is labeled S to indicate that it comes from the negative charge of the source of potential difference. This one electron S is repelled from the negative charge $-q$ at the left and is attracted by the positive charge $+q$ at the right. Therefore, the potential difference of the voltage source can make electron S move toward atom 1. Now atom 1 has an extra electron. As a result, the free electron of atom 1 can then move to atom 2. In this way, there is a drift of free electrons from atom to atom. The final result is that the one free electron labeled 8 at the extreme right in Fig. 1-6 moves out from the wire to return to the positive charge of the voltage source.

Considering this case of just one electron moving, note that the electron returning to the positive side of the voltage source is not the electron labeled S that left the negative side. All electrons are the same, however, and have the same charge. Therefore, the drift of free electrons resulted in the charge of one electron moving through the wire. This charge in motion is the current. With more electrons drifting through the wire, the charge of many electrons moves, resulting in more current.

The current through the wire has the speed of light, approximately, which is 297,600 km/s (186,000 mi/s). For ordinary electrical applications, where the wires are not long lines, we can consider that the

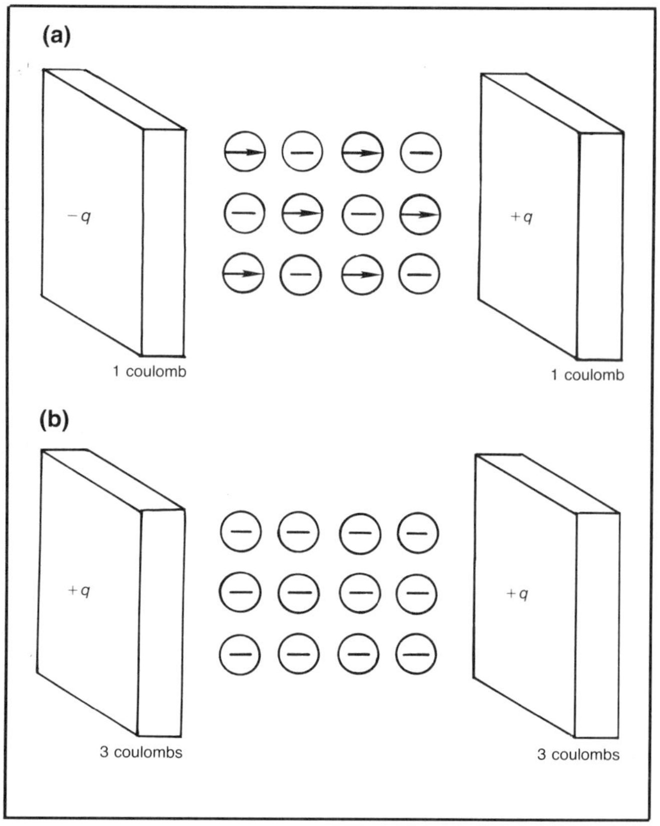

Figure 1-5. Work done in moving electrons between charges depend on their difference in potential. (a) Potential difference also equivalent to work by 2-coulomb charge. (b) Potential difference is zero between equal and opposite charges.

potential difference produces current instantaneously through the entire length of the wire. Furthermore, the current is the same at all points of the wire at the same time. Although a point nearer to the negative terminal of the voltage source has a greater repelling force on the free electrons, at this point the free electrons are farther from the positive terminal and have less attracting force. At the middle of the wire, the free electrons have equal forces of attraction and repulsion. Near the positive terminal, there is a greater attracting force on the free electrons but less repelling force from the negative terminal of the voltage source. In all cases, the total force causing motion of the free electrons is the same at any point of the wire, therefore resulting in the same current through all parts of the wire.

Potential Difference Is Necessary to Produce Current

The number of free electrons that can be forced to drift through the wire to produce the moving charge depends upon the amount of potential difference across the wire. With more applied voltage, the forces of attraction and repulsion can make more electrons drift, producing more charge in motion. A larger amount of charge moving with the same speed means a higher value of current. Less applied voltage across the same wire results in a smaller amount of charge in motion, which is a smaller value of current. With zero potential difference across the wire, there is no current.

The Ampere of Current

Current is movement of charge, therefore the unit for stating the amount of current is defined in rate of flow of charge. When the charge moves at the rate of 6.28×10^{18} electrons flowing past a given point per second, the value of the current is 1 ampere. This is the same as 1 coulomb of charge per second. Referring again to Fig. 1-6, note that if 6.28×10^{18} free electrons move past P_1 in a second, the current is 1 ampere. Similarly, the current is 1 ampere at P_2 because the electron drift is the same throughout the wire. If twice as many electrons moved past either point in 1, the current would be 2 amperes. The symbol for current is I, for intensity, because the current is a measure of how intense or concentrated the electron flow is. Sometimes current is called *amperage.*

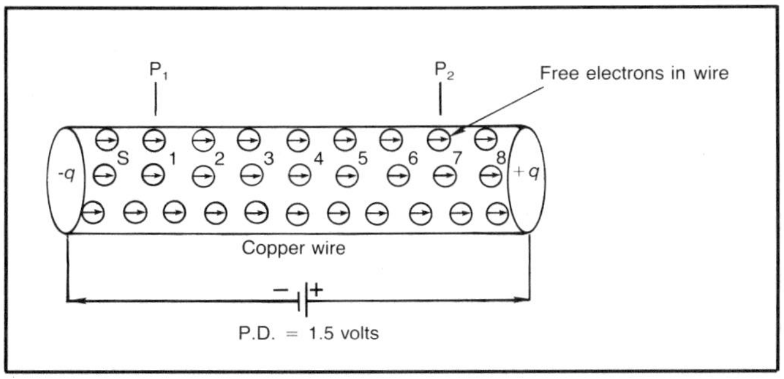

Figure 1-6. Potential difference (P.D.) across two ends of wire conductor causes drift of free electrons through the wire, producing electron current.

Resistance Is Opposition to Current

The fact that a wire conducting current can become hot is evidence of the fact that the work done by the applied voltage in producing current must be accomplished against some form of opposition. This opposition, which limits the amount of current that can be produced by the applied voltage, is called *resistance.* Conductors have very little resistance; insulators have a large amount of resistance.

The atoms of a copper wire have a large number of free electrons, which can be moved easily by a potential difference. Therefore, the copper wire has little opposition to the flow of free electrons when voltage is applied, corresponding to a low value of resistance. Carbon, however, has fewer free electrons than the copper. When the same amount of voltage is applied to the carbon as to the copper, far fewer electrons will flow because there are fewer free electrons. It should be noted that just as much current can be produced in the carbon by applying more voltage.

The Ohm of Resistance

The practical unit of resistance is the *ohm.* A resistance that develops 0.24 calorie* of heat when 1 ampere of current flows through it has 1 ohm of opposition. For example, a good conductor like copper wire can have a resistance of 0.01 ohm for a 1 ft length, the resistance-wire heating element in a 600-watt toaster has a resistance of 24 ohms, and the tungsten filament in a 100-watt light bulb has a resistance of 144 ohms. The abbreviation for resistance is R. The symbol for the ohm is the Greek letter omega (uppercase form, Ω). In diagrams, resistance is indicated by a zigzag line, as shown by R in Fig. 1-7.

Conductance

The opposite of resistance is conductance. The less the resistance, the higher the conductance. Its symbol is the Greek letter sigma (lowercase form, σ), and the unit is the mho, which is ohm spelled backward; σ is the reciprocal of R, or $\sigma = 1/R$. For example, 10 ohms of resistance is equal to 1/10 mho of conductance.

The Closed Circuit

In electrical applications requiring the use of current, the components are arranged in the form of a circuit, as shown in Fig. 1-7. A circuit can be

*One calorie is the quantity of heat that will raise the temperature of one gram of water by one degree centigrade.

defined as a path for current flow. The purpose of this circuit is to light the incandescent bulb. The bulb lights when the tungsten filament wire inside is white hot and produces an incandescent glow. By itself the tungsten filament cannot produce current. A source of potential difference is necessary. Because the dry cell produces a potential difference of 1.5 volts across the two output terminals, this voltage is connected across the filament of the bulb by means of the two wires so that the applied voltage can produce current through the filament.

In Fig. 1-7(b), the schematic diagram of the circuit is shown. Here the components are represented by shorthand symbols. The connecting wires are shown simply as straight lines because their resistance is small enough to be neglected. A resistance of less than 0.01 ohm for the wire is practically zero, compared with the 300-ohm resistance of the bulb. If the resistance of the wire must be considered, the schematic diagram includes it as additional resistance in the same current path. The schematic shows only the symbols for the components and their electrical connections.

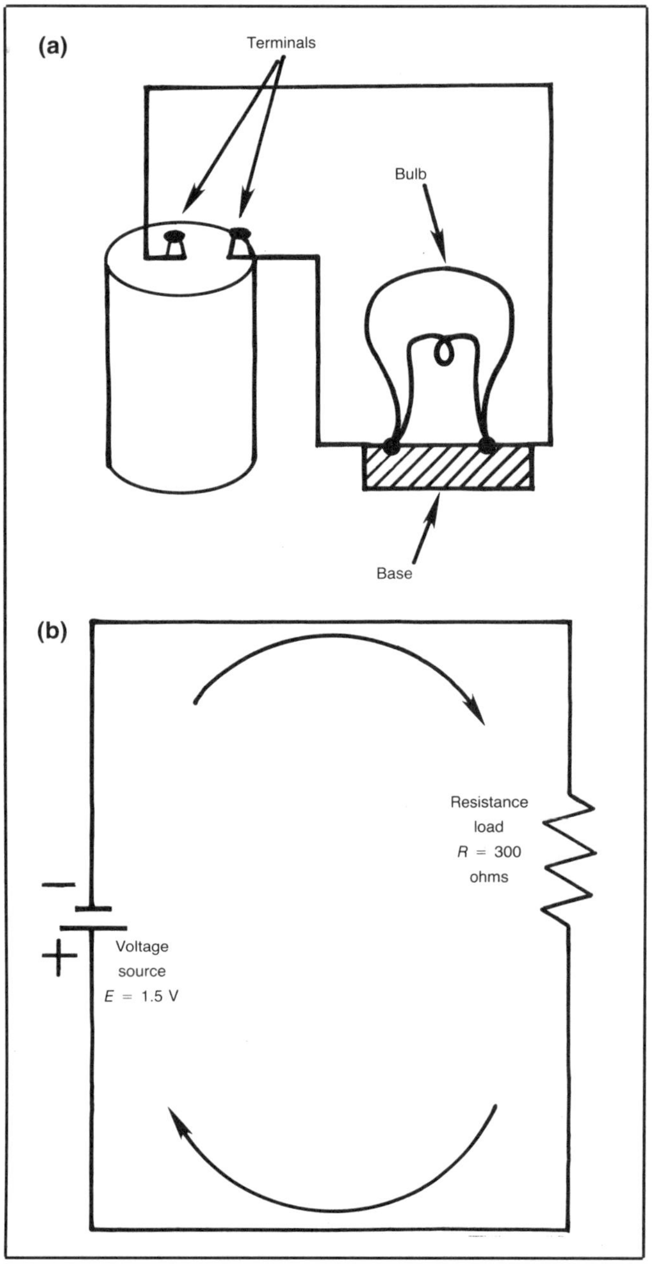

Figure 1-7. An electric circuit consisting of voltage source connected to resistance load. (a) Wiring diagram. (b) Schematic diagram.

Any electrical circuit has three important characteristics:

1. There must be a source of potential difference. Without the applied voltage, current cannot flow.
2. There must be a complete path for the current flow, from one side of the applied voltage source, through the external circuit, and back to the other side of the voltage source.
3. The current path normally has resistance. The resistance is in the circuit for the purpose of either generating heat or limiting the amount of current.

It is important to note that it is the current that moves through the circuit. The potential difference does not move. The voltage across the filament resistance makes the electrons flow from one side to the other, but the potential difference remains across the filament to do the work of moving electrons through the resistance of the filament.

As current flows in the circuit, electrons leave the negative (−) terminal of the cell, and the same number of free electrons in the conductor are returned to the positive (+) terminal. With electrons lost from the negative charge and gained by the positive charge, the two charges would tend to neutralize each other. The chemical action inside the dry cell, however, continuously separates electrons and protons to maintain the negative and positive charges on the outside terminals that provide the potential difference. Otherwise, the current would neutralize the charges, resulting in no potential difference, and the current would stop.

Direct Current

The current illustrated in Fig. 1-7(b) is direct current (DC) because it has just one direction. It has this one direction because the dry cell maintains the same polarity as output voltage. The cell is a DC voltage source; therefore it has one polarity of output voltage, which produces direct current in the external circuit. Alternating voltage periodically reverses in polarity, causing alternating current (AC) that periodically reverses in direction.

Table 1-2 summarizes the electrical characteristics of a circuit previously discussed.

Ohm's Law

The amount of current I in a circuit depends on its resistance R and the applied voltage E, as

Table 1-2
Electrical Characteristics

Characteristic	Symbol	Unit	Description
Charge	q	coulomb	Quantity of stored electrons or protons
Voltage	E	volt (V)	Potential difference between two unlike charges: makes electrons move
Current	I	ampere (A)	Electrons in motion
Resistance	R	ohm (Ω)	Opposition to flow of electrons that reduces amount of current
Conductance	σ	mho	Reciprocal of resistance

determined by the experimental law of George Ohm. If you know two of the factors E, I, and R, you can calculate the third.

The amount of current in a circuit, then, depends on these two factors: the voltage and the resistance. The voltage is applied across the resistance. Increasing the voltage produces a greater force, which tends to make more free electrons move around the circuit. Ohm's law can be stated specifically as

$$I = \frac{E}{R} \text{ or Amperes} = \frac{\text{volts}}{\text{ohms}} \qquad \text{(Eq. 1-1)}$$

where I is the current through the resistance R connected across the source of potential E.

Power

The unit of electrical power is the *watt*, named after James Watt. One watt of power equals the work done in 1 second by 1 volt of potential difference in moving 1 coulomb of charge. Because 1 coulomb per second is an ampere, power in watts equals the product of amperes times volts:

$$P = EI \text{ or Power (watts)} = \text{volts} \times \text{amperes} \qquad \text{(Eq. 1-2)}$$

For example, when a 6-volt battery produces 2 amperes in a circuit, the battery is generating 12 watts of power.

Power Dissipation in Resistance

When current flows in a resistance, heat is produced because the friction between the moving free electrons and the atoms obstructs the path of electron flow. The heat energy is evidence that power is used to produce current through a resistance. The power is generated by the source of applied voltage and consumed in the resistance in the form of heat. As much power as the resistance dissipates in heat must be supplied by the voltage source.

Because power is dissipated in the resistance of a circuit, it is convenient to express the power (P) in watts (W) in terms of resistance, as follows:

$$\text{For current,} \quad P = I^2R \qquad \text{(Eq. 1-3)}$$

$$\text{or, for voltage,} \quad P = E^2R \qquad \text{(Eq. 1-4)}$$

where

I is the current through a resistance,
R is the opposition in ohms, and
E is the potential difference in volts across R.

Chapter 2
Alternating Current and Voltage

Introduction

When voltage is induced across a loop of wire rotating through a magnetic field, the induced voltage alternates in polarity as the position of the loop reverses. The result is an alternating voltage, as shown in Fig. 2-1, which is a graph of the alternating voltage provided by the commercial 60-cycle AC power line. The complete waveform corresponds to four revolutions, or cycles, of the loop, which is rotating at the rate of 60 complete revolutions per second.

As the alternating voltage reverses in polarity, the current reverses in direction. The advantage of alternating current and voltage, as compared with direct current and voltage (discussed in Chapter 1), is not the reversing polarity so much as the fact that the value is continuously changing. This variation allows many applications that are not possible with a steady direct current. In fact, eddy current testing can be accomplished only by means of the varying field of alternating current and voltage. A steady current produces a constant magnetic field, but because the current does not vary, it (direct current) cannot produce electromagnetic induction.

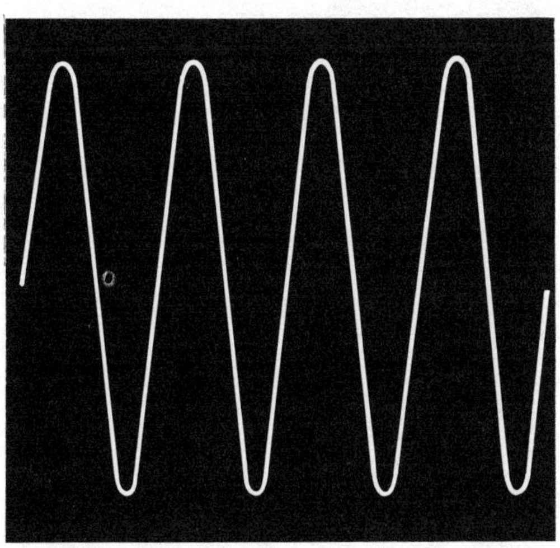

Figure 2-1. 60-cycle alternating voltage. Four cycles shown.

Fig. 2-2 illustrates a loop rotating through a magnetic field to generate the induced voltage *E* across its open terminals. The magnetic flux is vertical. In (a), the loop is shown in its horizontal starting position in a plane perpendicular to the magnetic field. When the loop rotates counterclockwise, the two longer conductors move around a circle. In the (a) position, motion of the loop does not induce a voltage because the conductors are perpendicular to the flux field. When the loop rotates through the parallel position in (b), the conductors cut across the flux, producing maximum induced voltage.

Each of the conductors has opposite polarity of induced voltage because the one at the top is moving to the left while the bottom conductor is moving to the right. The amount of voltage varies from zero to maximum as the loop moves from the flat position (a) to upright (b), where it can cut across the flux. Also the polarity at the terminals of the loop reverses as the motion of each conductor reverses during each half-revolution.

The Cycle

One complete revolution of the loop around the circle is a *cycle*. In Fig. 2-3, the loop is shown in its position at each quarter-turn during one complete

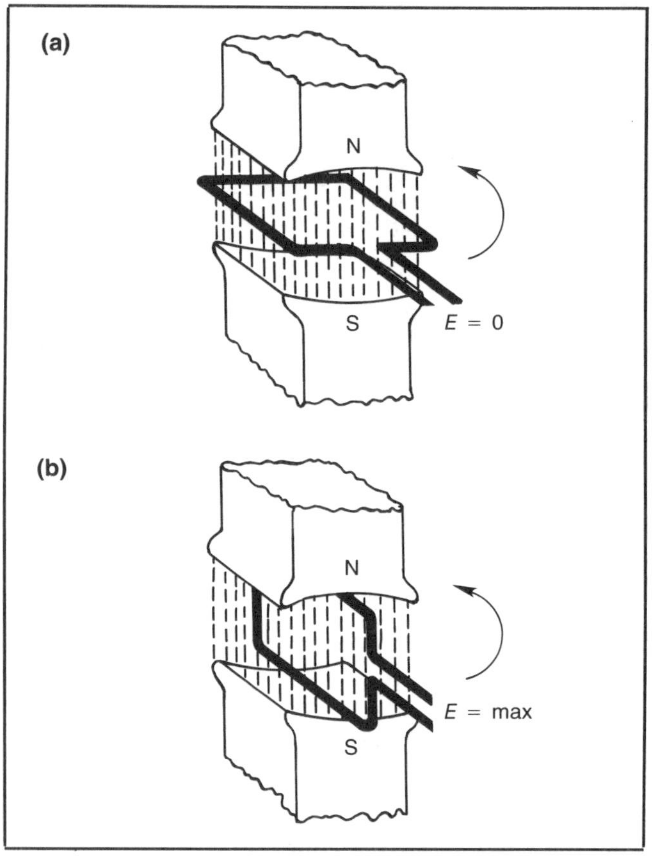

Figure 2-2. Loop rotating in a magnet field produces alternating induced voltage *E*. (a) Loop perpendicular to flux field, producing zero voltage. (b) Loop moving parallel to flux field, resulting in maximum induced voltage.

cycle. The corresponding wave of induced voltage also goes through one cycle.

At position A in Fig. 2-3, the loop is flat and perpendicular to the magnetic flux field, so that the induced voltage is zero. Counterclockwise rotation of the loop moves the dark conductor to the top position B, where it cuts across the field to produce maximum induced voltage. The polarity of the induced voltage here makes the dark conductor positive (+) and the light conductor negative (−).

In the graph of induced voltage values shown below the loop in Fig. 2-3, the polarity of the dark conductor is shown with respect to the other conductor. Positive voltage is shown above the horizontal axis on the graph. As the dark conductor rotates from its starting horizontal position toward the top vertical position, where it cuts the maximum flux, more and more voltage is induced, with positive polarity. When the loop rotates through the next quarter-turn, it returns to the horizontal position shown in position C, where it cannot cut across the flux field. Therefore, the graph of induced voltage decreases from its maximum value to zero at the half-turn.

The next quarter-turn of the loop moves it to the position shown at position D in Fig. 2-3 where the loop cuts across the flux again for maximum induced voltage. Note, however, that here the dark conductor is moving left to right at the bottom of the loop. This motion is reversed from the direction it had when it was at the top. Because of the reversed direction of motion during the second half-revolution, the induced voltage has opposite

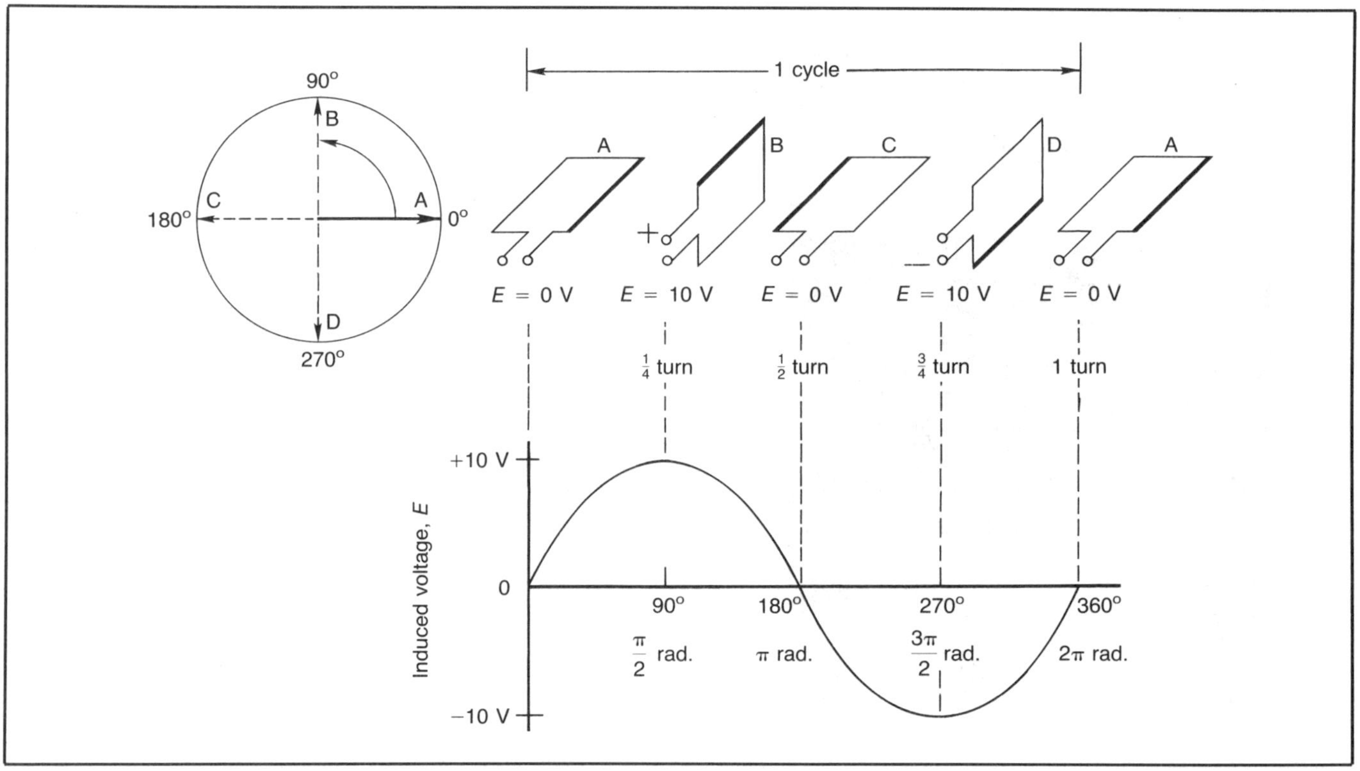

Figure 2-3. One cycle of sine-wave alternating voltage generated by the loop.

polarity, with the dark conductor negative. This polarity is shown in the graph as negative voltage, below the horizontal axis.

Because the cycle of voltage in Fig. 2-3 corresponds to rotation of the loop around a circle, it is convenient to consider parts of the cycle in angles. The complete circle includes 360 degrees. One half-cycle, or one alternation, is 180 degrees of revolution, and a quarter-turn is 90 degrees. The circle next to the loop positions in Fig. 2-3 illustrates the angular rotation of the dark conductor counterclockwise from 0 degrees to 90 degrees to 180 degrees for one half-cycle, and then to 270 degrees, returning to 360 degrees to complete the cycle.

In angular measure it is convenient to use a specific unit angle called the *radian,* which is an angle equal to 57.3 degrees. Its convenience is due to the fact that a radian is the angular part of the circle that includes an arc equal to the radius r of the circle, as shown in Fig. 2-4. Because the circumference around the circle equals $2\pi r$, a circle includes 2π radians. Therefore, one cycle equals 2π radians.

As shown in the graph in Fig. 2-3, divisions of the cycle can be indicated by angles either in degrees or radians. Zero degrees is also zero radians, 360 degrees is 2π radians, 180 degrees is π radians, 90 degrees is $\pi/2$ radians, and 270 degrees is π radians plus $\pi/2$ radians, which equals $3\pi/2$ radians. The constant 2π in circular measure is numerically equal to 6.28.

The voltage waveform in Figs. 2-1 and 2-3 is called a sine wave, sinusoidal wave, or sinusoid because the amount of induced voltage is proportional to the *sine* of the angle of rotation in circular motion producing the voltage. The sine is a trigonometric function of an angle. The instantaneous value of a sine-wave voltage for any angle of rotation is expressed by the formula

$$e = E \sin \theta \text{ (volts)} \qquad \text{(Eq. 2-1)}$$

where θ (Greek letter theta) is the angle, sin is the abbreviation for its sine, E is the maximum voltage value, and e is the instantaneous value for any angle.

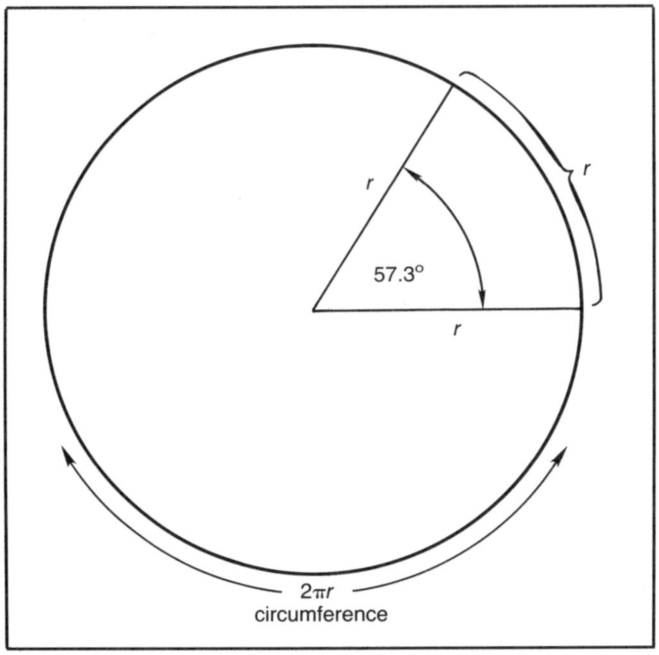

Figure 2-4. One radian is the angle equal to 57.3 degrees. The complete circle includes 2π radians.

Frequency

The number of cycles per second (cps) is the frequency. By definition, one cps is equal to one hertz (Hz). In Fig. 2-2, if the loop rotates through 60 complete revolutions, or cycles, during one second, the frequency of the generated voltage is 60 cps or 60 Hz.

Note that the factor of time is involved. More cycles per second means a higher frequency and less time for one cycle, as illustrated in Fig. 2-5. A complete cycle is measured between two successive points that have the same value and direction. On the time scale of one second, the waveform in (a) goes through one cycle, for a frequency of 1 Hz; the waveform in (b) has much faster variations, with four complete cycles during one second, or a frequency of 4 Hz. Both waveforms are sine waves, even though each has a different frequency.

At the present time, eddy current instruments usually operate in a range from 60 Hz (60 cps) to 6 MHz (6 million cps).

For high frequencies, kilocycles per second and megacycles per second are commonly used:

One kilocycle per second = 1 kHz
= 1,000 Hz
= 1×10^3 Hz

One megacycle per second = 1 MHz
= 1,000,000 Hz
= 1×10^6 Hz

Period

The amount of time for one cycle is the *period.* Its symbol is T, for time. With a frequency of 60 Hz, for example, the time for one cycle is 1/60 s. Therefore the period is 1/60 s in this case.

The frequency and period are reciprocals of each other:

$$T \text{ (s)} = \frac{1}{f \text{ (Hz)}} \text{ or } f \text{ (Hz)} = \frac{1}{T \text{ (s)}} \qquad \text{(Eq. 2-2)}$$

where time T is in seconds and the frequency f (sometimes represented by ν, the Greek letter nu, lowercase form) is in hertz.

The higher the frequency, the shorter is the period. In Fig. 2-5, the period for the wave in (a), with a frequency of 1 Hz, is 1 s; the higher-frequency wave of 4 Hz in (b) has a period of 1/4 s for a complete cycle.

Wavelength

When a periodic variation is considered with respect to distance, one *wavelength* is the length of one complete cycle (Fig. 2-6). The distance traveled by the wave in one cycle is the wavelength. The wavelength depends upon the frequency of the variation and its velocity of transmission:

$$\lambda = \frac{\text{velocity}}{\text{frequency}} \text{ or } \lambda = \frac{c}{f} \qquad \text{(Eq. 2-3)}$$

where λ (the Greek letter lambda, lowercase form), is the symbol for one complete wavelength. For electromagnetic eddy current waves, the velocity in air or vacuum is 3×10^{10} cm/s, which is the speed of light. Therefore,

$$\lambda = \frac{(3 \times 10^{10} \text{ cm/s})}{f \text{ (Hz)}} \qquad \text{(Eq. 2-4)}$$

where wavelength is in centimeters, frequency is in hertz.

Note that the higher the frequency is, the shorter the wavelength.

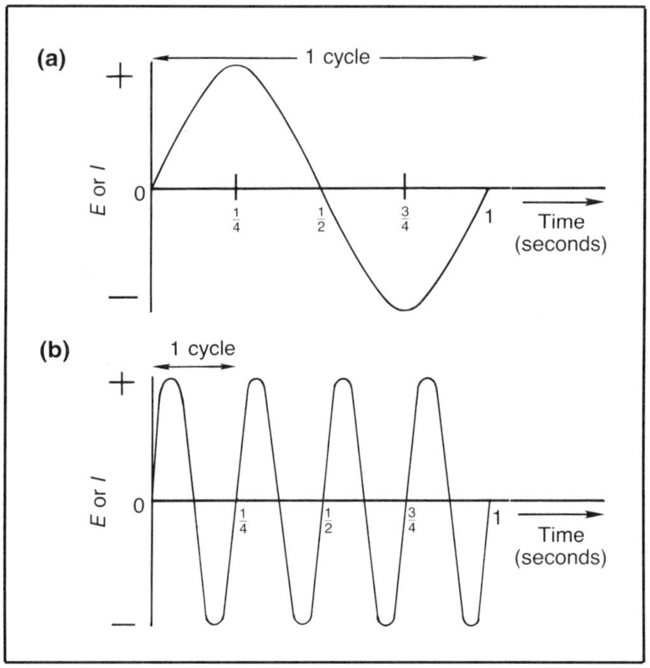

Figure 2-5. Number of cycles per second is the frequency in hertz (Hz). (a) Frequency of 1 Hz. (b) Frequency of 4 Hz.

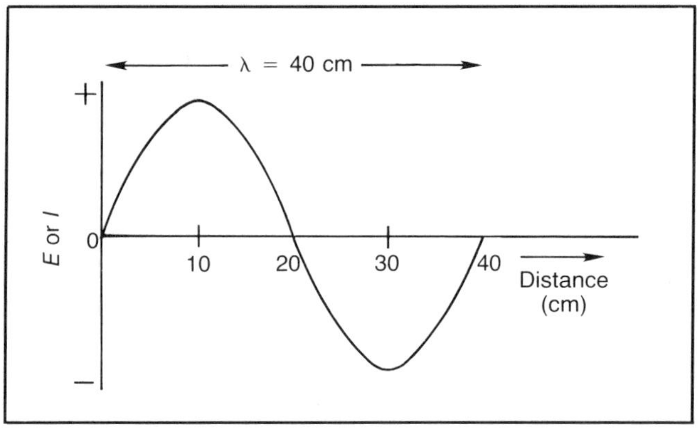

Figure 2-6. Wavelength (**λ**) is the distance traveled by the wave in one cycle. Here the wavelength is 40 cm.

Phase Angle

In Fig. 2-7, wave B starts at maximum, and wave A starts at zero. The complete cycle of wave B through 360 degrees takes it to the maximum value from which it started. Wave A starts and finishes its cycle at zero. With respect to time, therefore, wave B is ahead of wave A in its values of generated voltage. The amount it leads in time equals one quarter-revolution, which is 90 degrees. This angular difference is the phase angle between waves B and A. Wave B leads wave A by the phase angle of 90 degrees.

Two waves can be out of phase by any angle less or more than 90 degrees. A phase angle of 0 degrees, however, means the two waves are *in phase.* A phase angle of 180 degrees corresponds to opposite phase for the two waves, and waves of equal and opposite phase cancel each other. The waves must have the same frequency or they cannot be compared in phase.

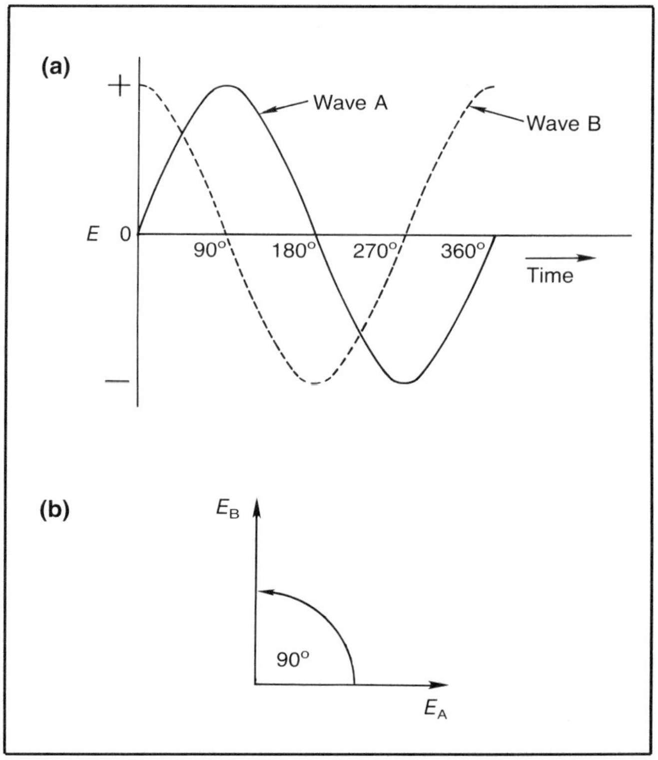

Figure 2-7. Two sine-wave voltages 90 degrees out of phase. (a) Sine-wave voltage for wave B leads voltage for wave A by 90 degrees. (b) Phase angle of 90 degrees between vectors E_B and E_A representing the two voltage waves.

Phase-Angle Vectors

In comparisons of the phase of alternating currents and voltages, it is much more convenient to use vector diagrams corresponding to the voltage and current waveforms, as shown in Fig. 2-7(b). The arrows represent the vector quantities of generated voltage. A vector is a quantity that has magnitude and direction. The length of the arrow indicates the magnitude of the alternating voltage. The angle of the arrow with respect to the horizontal axis indicates the phase angle. Note that the leading voltage vector E_B in Fig. 2-7(b) is shown 90 degrees counterclockwise from E_A. This arrangement is standard practice; counterclockwise rotation is considered the positive direction for angles.

Chapter 3
Magnetism

Introduction

Electrical energy exists in two forms, voltage and current. In terms of voltage, separated electric charges have the potential to do mechanical work in attracting or repelling charges. Similarly, with electron flow the current has an associated magnetic force called magnetism. The magnetism is an invisible force that has the ability to do mechanical work of attraction or repulsion. Materials made of iron, nickel, and cobalt particularly concentrate their magnetic effect at opposite ends. As shown in Fig. 3-1, these points of concentrated magnetic strength are defined as north and south poles. Because the magnetic poles are opposite in the same sense as positive and negative polarities, there is a mechanical force of attraction between unlike poles and repulsion between like poles.

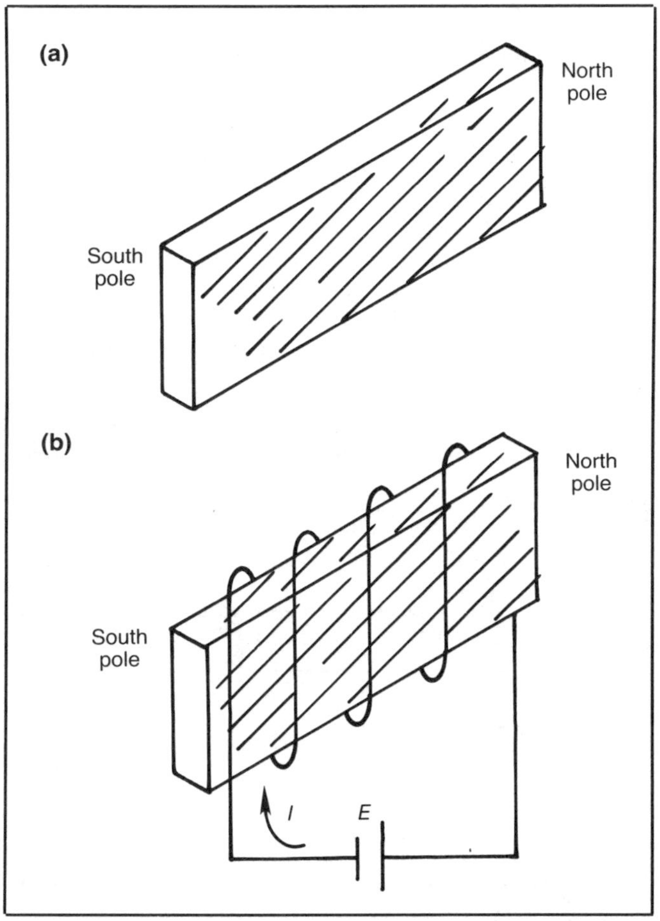

Figure 3-1. Poles of a magnet. (a) Permanent magnet. (b) Electromagnet.

When the magnetic poles are present without any external current, as in (a), this forms a permanent magnet. When the magnetic poles are produced by current from an external voltage source, as in (b), this forms an electromagnet.

Molecular Magnets

If a permanent bar magnet is cut in two, each half becomes a magnet itself. Both magnets have a north and south pole and each one has half the original pole strength. This is explained by the assumption that magnetic materials contain tiny molecular magnets called *magnetic dipoles.* When the dipoles are in line, all pointing the same way, as in Fig. 3-2(a), the molecular magnets reinforce each other to produce strong magnetic poles at the opposite ends of the bar magnet. When the dipoles have different directions, as in Fig. 3-2(b), they cancel and the bar does not have magnetic poles.

The internal magnetic dipoles may be considered as the magnetic force associated with moving electrons within each atom. Where the atomic structure allows the electrons to move in similar directions, the current has a net magnetic force that provides magnetic dipoles. In nonmagnetic materials, however, the electrons within each atom move in opposite directions, canceling the magnetic effects of the electrons, and the substance does not have magnetic dipoles.

Magnetic Field

The space outside a magnet where its poles have a force of attraction or repulsion is a magnetic *field of force*. The field exists in all directions around the magnet.

It is the field of a magnet that enables it to induce magnetism in an iron bar, without the two touching, as long as the iron is in a part of the field strong enough to line up its molecular dipoles.

The magnetic field is invisible, but evidence of its force can be seen when small iron filings are sprinkled on a glass or paper sheet over a bar magnet, as illustrated in Fig. 3-3(a). Each iron filing becomes a small magnet by induction. Many filings cling to the poles of the magnet, showing that the field is strongest at the poles.

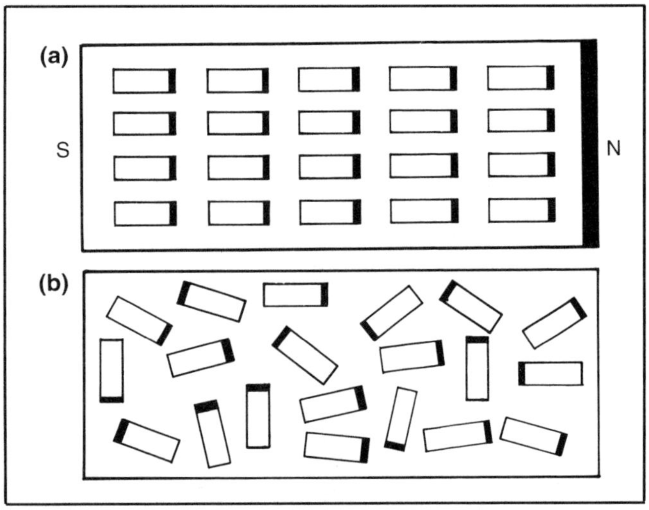

Figure 3-2. Molecular dipole magnets in iron bar. (a) Dipoles in line when iron is magnetized. (b) Disorganized dipoles canceling pole strength when iron is not magnetized.

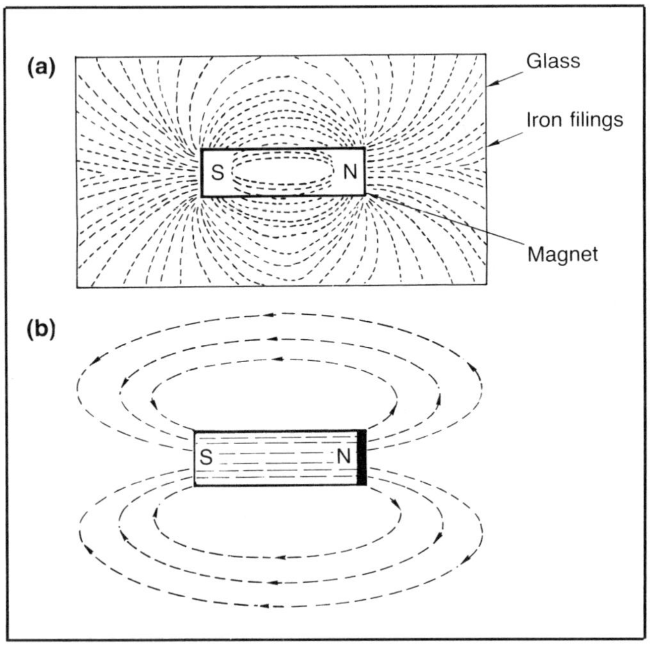

Figure 3-3. Magnetic field of force around bar magnet. (a) Field outlined by iron filings. (b) Field indicated by lines of force.

Magnetic Flux

The entire group of magnetic field lines, which can be considered to flow outward from the north pole of a magnet, is called *magnetic flux*. Its symbol is the Greek letter phi, ϕ (lowercase form). The unit of flux is the maxwell (named for James Clerk Maxwell) in the metric system. One maxwell is equal to one magnetic field line. In Fig. 3-4, for example, the flux illustrated is equal to 6 maxwells because there are six lines flowing into or out of each pole.

Flux Density

As shown in Fig. 3-4, the *flux density* is the number of magnetic field lines or maxwells per unit area in a section perpendicular to the direction of flux. Its symbol is B. The unit is lines per square inch in the English system, or lines per square centimeter in the metric system. One line per square centimeter is called a gauss (named for Karl Gauss). In Fig. 3-4, the total flux ϕ of six lines, or 6 maxwells, has a flux density B of 2 gauss at point P in the field because there are two lines per cm^2. The flux density has a higher value close to the poles, where the flux lines are more dense.

Permeability

The ability of a material to concentrate magnetic lines is called permeability. Any material that is easily magnetized, such as soft iron, concentrates the magnetic flux. Numerical values of permeability for different materials are assigned in comparison with air or vacuum. Because air, vacuum, or any

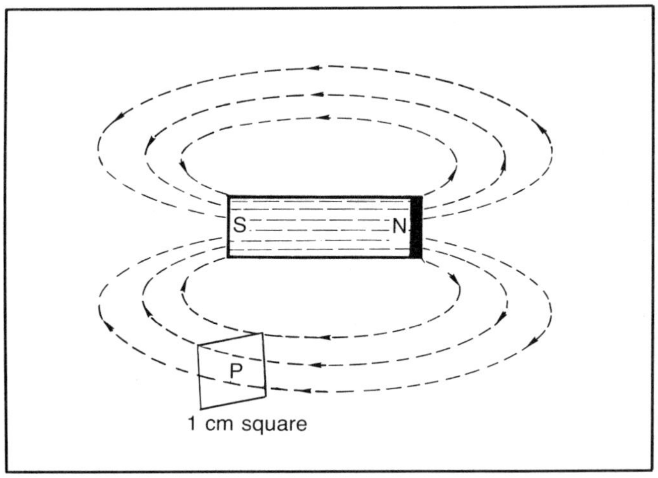

Figure 3-4. Total magnetic flux of six lines, equal to 6 maxwells. At P the flux density is two lines per cm^2, equal to 2 gauss.

nonmagnetic material cannot affect a magnetic field by induction, they all have the reference value of 1 for permeability. The symbol for permeability is μ (Greek mu, lowercase form). Typical values of μ for iron are 100 to 5,000. There are no units for permeability (dimensionless) because it is a ratio comparing two flux densities.

Magnetizing Force

The ability of a magnetic field to produce magnetism by induction is its *magnetizing force*, or *field intensity*, H. Its unit is the oersted (named for H. C. Oersted). When the magnetic field exerts a force of 1 dyne upon a standard north pole, the field intensity, or magnetizing force, is 1 oersted at that point. The dyne is a basic unit of mechanical force in the metric system. The force can be either attraction or repulsion. The field intensity decreases inversely as the square of the distance from the poles.

The relation between the magnetizing force H in oersted units, flux density B in gauss, and permeability is

$$B = \mu H \qquad \text{(Eq. 3-1)}$$

With a material of high μ in the field, H produces a large value of B. For example, for iron with a μ of 50 in a field having the intensity H of 100 oersteds, the flux density B in the iron equals 5,000 gauss. In air, with μ equal to 1, the flux density would be 100 gauss.

B-H Magnetization Curve

A graph may be used to indicate the flux density B for different amounts of magnetizing force H. A typical B-H magnetizing curve is shown in Fig. 3-5 for soft iron. The horizontal base line, or X axis, is marked off in units of magnetizing force H, in oersteds. The vertical, or Y, axis indicates flux density B in gauss.

Different materials have different curves on the B-H graph. The effect of little increase in flux density B when the magnetizing force H is increased is called saturation. The part of the B-H curve where saturation begins is the "knee" of the curve.

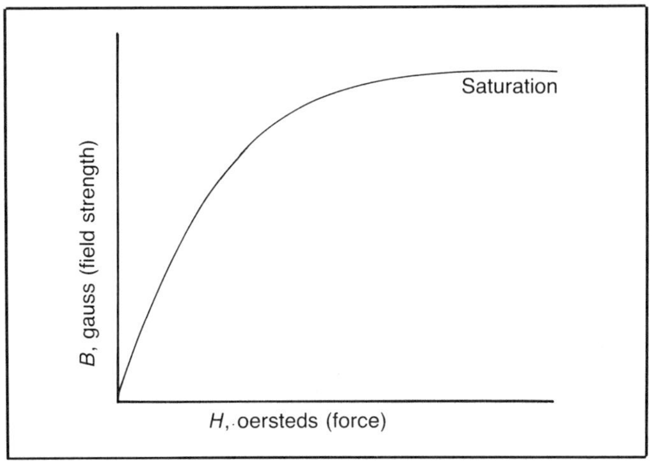

Figure 3-5. *B-H* magnetization for soft iron.

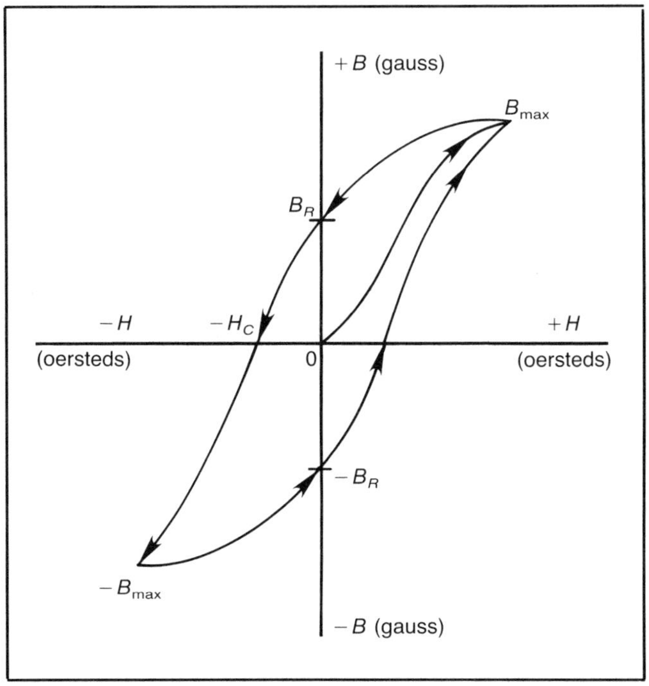

Figure 3-6. Hysteresis loop. This is a *B-H* curve with alternating magnetizing force *H*.

Magnetic Hysteresis

Hysteresis means "a lagging behind." With respect to the magnetic flux in an iron core of an electromagnet, the flux lags the increases or decreases of magnetizing force. The hysteresis results from the fact that the magnetic dipoles are not perfectly elastic and do not return to their original positions when the external force is removed.

When the magnetizing force varies at a slow rate, the hysteresis effect can be considered negligible. The faster the magnetizing force changes, however, the greater is the hysteresis effect and loss of energy. The work done by the magnetizing force against this internal friction produces heat. This energy wasted in heat as the molecular dipoles lag the magnetizing force is called hysteresis loss.

For steel and other hard magnetic materials, the hysteresis losses are much higher than in soft magnetic materials, such as iron or ferrites. To show the hysteresis characteristics of a substance, its values of flux densities are plotted on a B-H graph, as in Fig. 3-6, with periodically reversing values of magnetizing force. The curve is the hysteresis loop, and the slope of the curve is the changing μ of the material.

The larger the loop, the greater the hysteresis loss is. The curve starts from zero, the origin, when the material is unmagnetized, rises to saturation at $+B_{max}$ with a positive direction of magnetizing field, then traces the maximum values from $+B_R$ through $-B_{max}$ with a negative direction (to the left) for H, and back to $+B_{max}$ again.

The value of B_R, which is the flux density remaining after the magnetizing force has been reduced to zero, is the *residual induction* (or *residual magnetism*, or *remanence*) of a magnetic material, also called *retentivity*. The value of $-H_C$ in Fig. 3-6, which equals the magnetizing force that must be applied in the reverse direction to reduce the flux to zero, is the *coercive force*, or *reluctance*, of the material.

To demagnetize a material completely, the residual induction must be reduced to zero. The easiest method is to magnetize and demagnetize the material, following the form of the hysteresis loop in Fig. 3-6, by means of an alternating field. Then, as the material is gradually withdrawn from the field, or the field from the material, the hysteresis loop becomes smaller and smaller. Finally, with the weakest field, the loop collapses to practically zero, resulting in zero residual induction. For this reason, alternating current is generally used for demagnetization.

Chapter 4
Electromagnetism

Introduction

The link between magnetism and electricity was discovered in 1824 by Oersted, who found that current in a wire could move a magnetic compass needle outside the wire. A few years later the opposite effect was discovered: a magnetic field in motion can force electrons to move, producing current. This important discovery was made in 1831 independently by Faraday and Henry. Electromagnetism, therefore, consists of the magnetic effects of electrical current. Electrons in motion have an associated magnetic field; a moving magnetic field can produce current. These electromagnetic effects have many applications that are the basis of eddy current testing.

A Magnetic Field about a Conductor

The fact that there is a magnetic field associated with current flowing in a wire is illustrated in Fig. 4-1, where iron filings are attracted to the wire when the current is flowing and are not attracted when the current is stopped.

The magnetic field is strongest at the surface of the conductor, decreasing inversely as the square of the distance from the conductor.

With circular lines of magnetic force, a north pole in the field would tend to move in a circular path. The direction must be considered in terms of clockwise or counterclockwise rotation. In Fig. 4-2, the compass needle shows a north pole would move counterclockwise in the field. To determine the circular direction of the field, the following rule can be used: if you look down the wire in the direction of electron flow, the magnetic field is counterclockwise, as shown in Fig. 4-2.

Magnetic Polarity of a Coil

Bending a straight conductor in the form of a loop, as shown in Fig. 4-3, has two effects. First, the magnetic field lines are more dense inside the loop. The total number of lines is the same as for the straight conductor, but in the loop the lines are concentrated in a smaller space. Second, inside the loop, all the lines are adding because they are traveling in the same direction. This makes the loop field effectively the same as a bar magnet, with opposite poles at opposite faces of the loop.

A coil of conducting wire with more than one turn is generally called a *solenoid.* An ideal solenoid, however, has a length much greater than its diameter. Like a single loop, the solenoid concentrates the magnetic field inside the coil and provides opposite magnetic poles at the ends. These effects are multiplied, however, by the number of turns. Referring to Fig. 4-4(a), note that

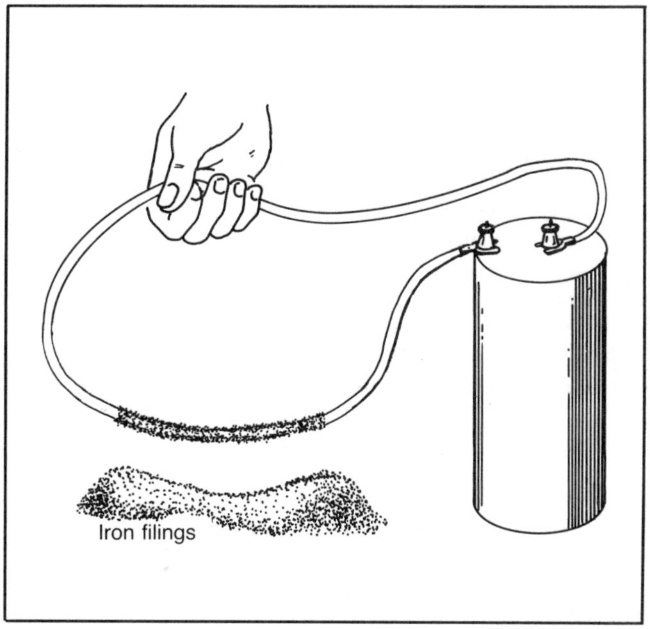

Figure 4-1. Iron filings attracted by magnetic field of conductor carrying current.

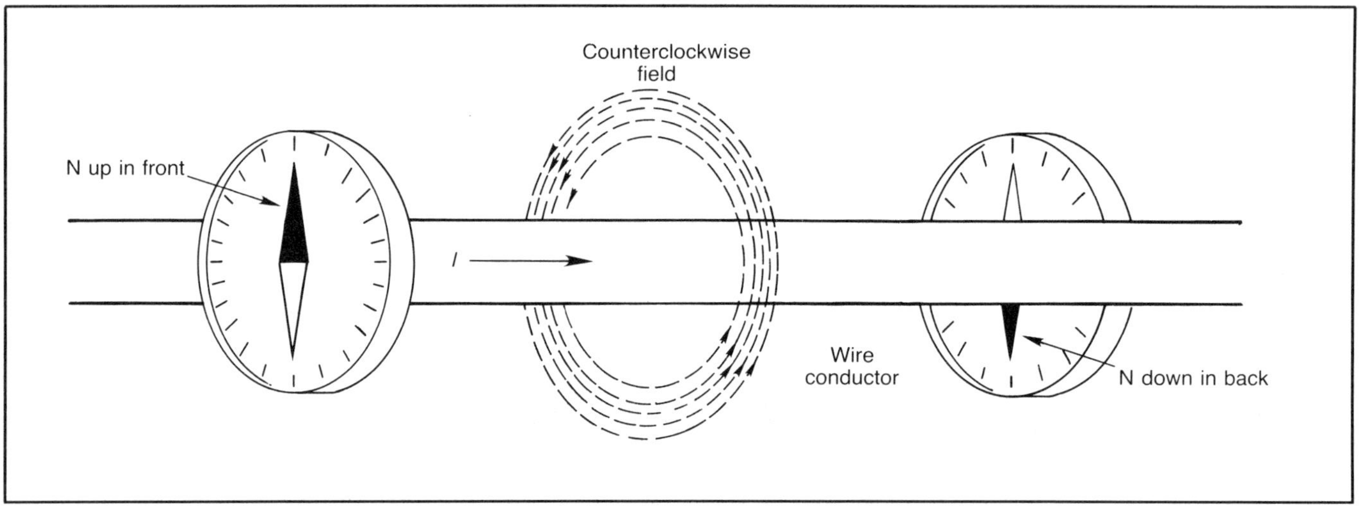

Figure 4-2. Counterclockwise field around a straight conductor.

the magnetic field lines aid each other in the same direction inside the coil. The field is strongest at the center. Outside the coil, the field corresponds to that of a bar magnet, with north and south poles at opposite ends of the solenoid, as illustrated in Fig. 4-4(b).

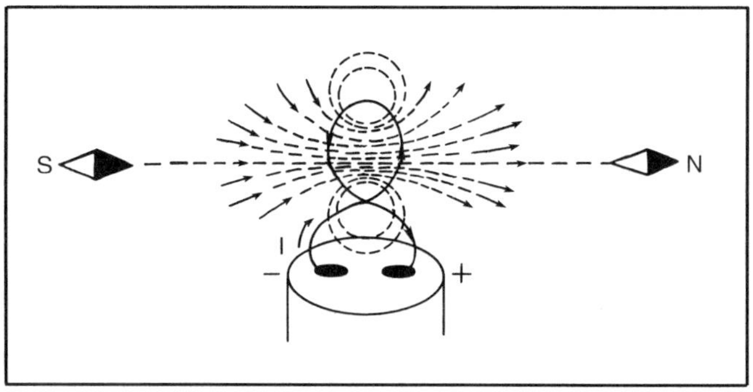

Figure 4-3. Magnetic poles of a current loop.

Magnetic Induction

Just as electrons in motion have an associated magnetic field, so, when a magnetic flux moves, the motion of magnetic lines cutting across a conductor forces free electrons in the conductor to move, producing current. The process is called *induction* because there is no physical connection between the magnet and the conductor. The induced current is a result of generator action as the mechanical work put into moving the magnetic field is converted into electrical energy when current flows in the conductor. Without motion, there is no current.

The motion is necessary to have the lines of the magnetic field cut across the conductor. This cutting can be accomplished by motion of either the field or the conductor.

To have electromagnetic induction, the conductor and the magnetic lines of flux must be perpendicular (normal) to each other so that motion makes the flux cut through the cross-sectional area of the conductor. As shown in Fig. 4-5, the conductor is at right angles to the lines of force in the field H. The reason these must be perpendicular is to make the induced current have an associated magnetic field in the same plane as the external flux. When the magnet is moved downward, current flow is in the direction shown (A toward B). If the magnet is moved upward, current will flow in the opposite direction.

Consider the case of magnetic flux cutting a conductor that is not a closed circuit, as shown in Fig. 4-6. The motion of the flux across the conductor forces free electrons to move, but, with an open circuit, the displaced electrons produce electric charges at the two open ends. For the direction shown, free electrons in the conductor are forced to move to point A, and electrons accumulate there. Point A then develops a negative potential. At the same time, point B loses electrons and becomes positively charged. The result is a potential difference across the two ends, provided by the separation of electric charges in the conductor. The potential difference is an electromotive force, generated by the work of moving the flux.

The amount of induced voltage produced by flux cutting the turns of a coil depends upon the following three factors:

1. *Amount of flux.* The more magnetic lines of force cutting a conductor, the higher the induced voltage is.
2. *Time rate of cutting.* The faster the flux cuts a conductor, the higher the induced voltage is.
3. *Number of turns.* The more turns of a coil, the higher its induced voltage is.

The induced voltage can be calculated in volts from the formula

$$E_{ind} = \frac{(N \cdot \theta)}{T \cdot 10^8} \text{ volts} \qquad \text{(Eq. 4-1)}$$

where

θ = maxwells or number of lines,
T = time in seconds, and
N = number of turns.

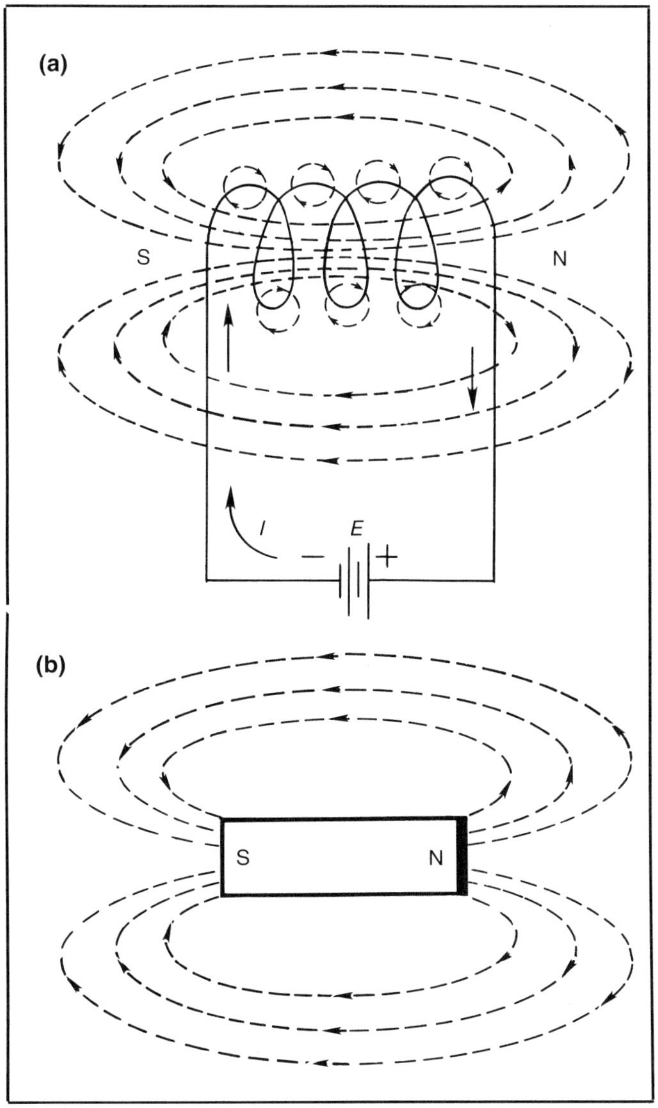

Figure 4-4. Magnetic poles of a solenoid. (a) Coil winding. (b) Equivalent bar magnet.

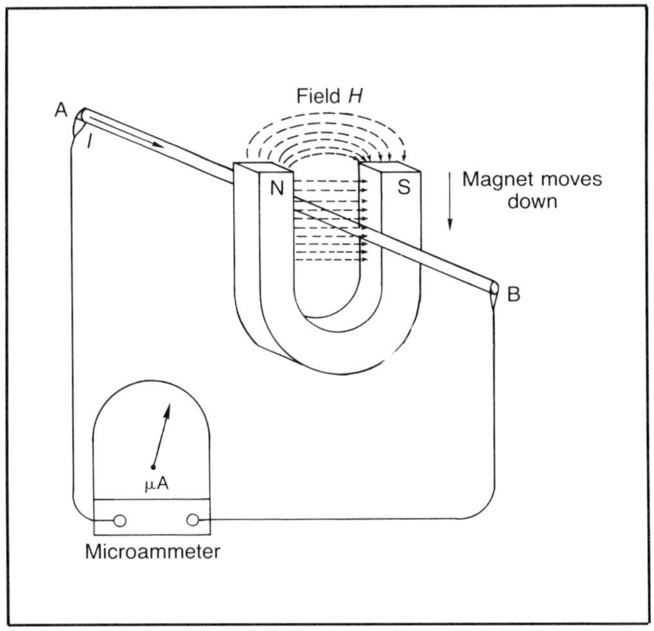

Figure 4-5. Induced current produced by magnetic flux cutting across a conductor.

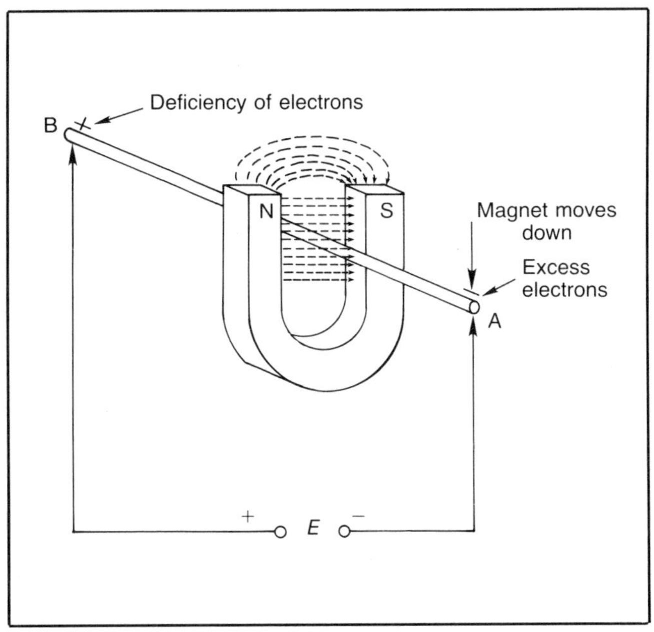

Figure 4-6. Voltage induced across open ends of conductor cut by magnetic flux in motion.

Chapter 5
Induction by Alternating Current

Induction

Induced voltage is the result of a magnetic flux cutting across a conductor, produced by physical motion of either the magnetic field or the conductor. When the current in a conductor varies in amplitude, however, the variations of current and its associated magnetic field are equivalent to motion of the flux. As the current increases in value, the magnetic field expands outward from the conductor. When the current decreases, the field collapses into the conductor. As the field expands and collapses with changes of current, the flux is effectively in motion. Therefore, a varying current can produce *induced voltage* without the need for motion of the conductor.

Fig. 5-1 illustrates the changes in magnetic field associated with a sine wave of alternating current. Because the alternating current varies in amplitude and reverses in direction, its associated magnetic field has the same variations.

The result of an expanding and collapsing flux field is the same as that of a field in motion. This moving flux cuts across the conductor that is providing the current, producing induced voltage in the wire itself. Furthermore, any other conductor in the field, whether carrying current or not, also is cut by the varying flux and has induced voltage. The *inductance* is an additional characteristic of the circuit besides its resistance.

Lenz's Law

This law states that any voltage generated by induction must oppose the motion producing the induced voltage. In terms of induced voltage produced by varying current, the change of current is equivalent to motion of the magnetic flux and must be opposed by the induced voltage. When the current increases, the induced voltage opposes the decrease. In both cases, the change is opposed by the induced voltage. *Inductance,* therefore, is the characteristic that opposes any change in current. An induced voltage is often called a back emf, or counter emf (electromotive force).

Self-Inductance

The ability of a conductor to induce voltage in itself when the current changes is its *self-inductance,* or simply *inductance.* The symbol for inductance is *L*; its unit is the henry. One henry is the amount of inductance that allows one volt to be induced when the current changes at the rate of one ampere per second. The formula is

$$L = \frac{-E_{\text{ind}}}{(\Delta I/T)} \text{ or } 1 \text{ henry} = \frac{1 \text{ volt}}{1 \text{ A/s}} \qquad \text{(Eq. 5-1)}$$

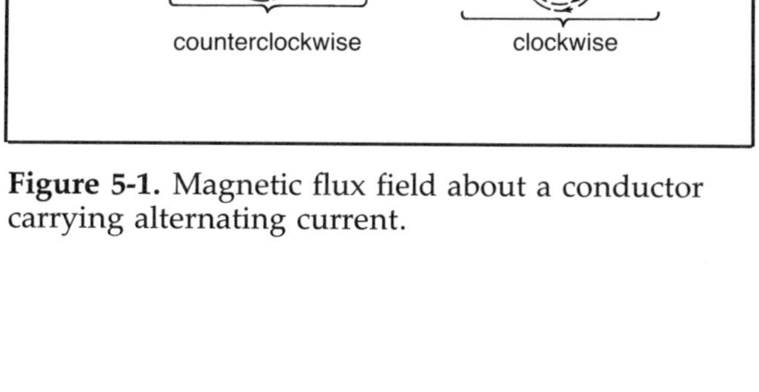

Figure 5-1. Magnetic flux field about a conductor carrying alternating current.

where

L = inductance,
E_{ind} = induced voltage,
Δ = change in current, and
T = time.

E is in volts and $\Delta I/T$ is the current change in amperes per second. The negative sign for E indicates that the polarity of induced voltage is in opposition to the current change, but the polarity can be disregarded in calculating the value of L.

The inductance of a coil increases with the (1) number of turns, (2) diameter of the coil, and (3) permeability of the core.

For a straight air-core coil, the inductance increases as the square of the turns and diameter. Doubling the turns provides four times the inductance, and doubling the diameter provides four times the inductance. If both the number of turns and the diameter are doubled, the inductance is increased by a factor of 16. The inductance increases directly with the length; i.e., doubling the length provides twice the inductance.

Mutual Inductance

When the current in an inductor changes, the varying flux can cut across any other inductor nearby, producing induced voltage in both inductors, as illustrated in Fig. 5-2. The coil L_1 is connected to a generator that produces varying current ΔI in the turns. The winding L_2 is not connected to L_1, but the turns are linked by the magnetic field. Varying the current in L_1 therefore induces voltage across L_1 and across L_2.

If all of the flux of current in L_1 links all the turns of the coil L_2, each turn in L_2 will have the same amount of induced voltage as each turn in L_1.

Additionally, when the induced voltage produces current in L_2, its varying magnetic field induces voltage in L_1. The two coils have mutual inductance because current in one coil can induce voltage in the other.

In eddy current testing, alternating current is supplied to the sensing coil. This alternating current also produces an alternating magnetic field which, when placed near a metal conductor, causes current to flow in the metal conductor by mutual inductance. The current in the conductor (eddy current) will generate a secondary magnetic field, which induces a current in the sensor coil. This mutual inductance causes a change in the impedance of the coil. The impedance signals sensed by the search coil are the measurements of the test specimen. Hence, the eddy current technique uses the effect of electromagnetic fields and induction to characterize physical properties of metallic materials.

Fig. 5-3 illustrates the generation of eddy currents in the conductive test specimen by mutual induction between the coil's magnetic field H_0 and the specimen's reaction magnetic field H_r.

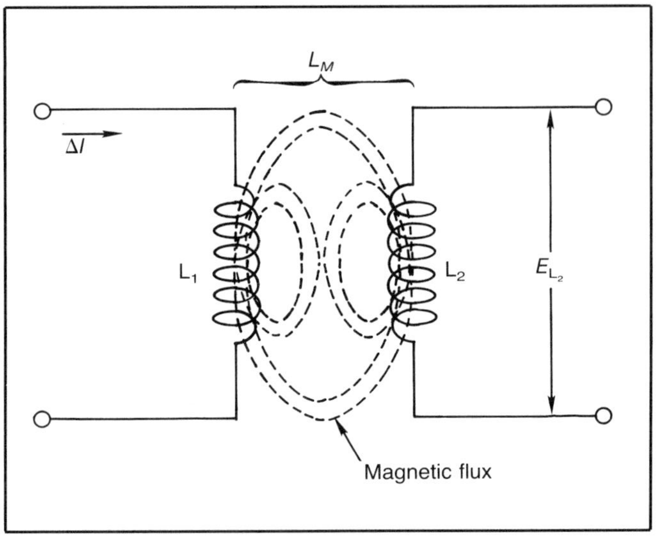

Figure 5-2. Mutual inductance L_M between two coils L_1 and L_2 linked by magnetic flux.

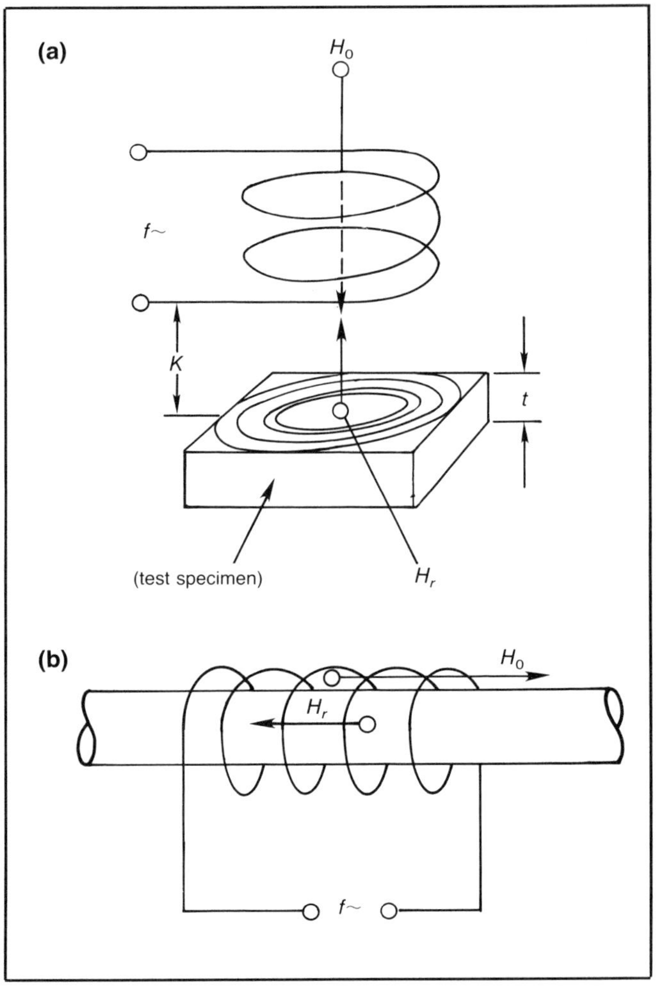

Figure 5-3. Mutual inductance between the sensing coils' magnetic field H_0 and the conductive test specimen's magnetic field H_r. (a) Probe coil above flat surface of metallic test object. (b) Test object within cylindrical test coil.

Coefficient of Coupling *(k)*

The fraction of total flux from one coil linking another coil is the *coefficient of coupling k.* The coefficient increases by placing the sensing coil close to the conductive specimen. When probe coils are used, the spacing between the coil and conductor is called *liftoff;* when encircling coils are used, the coupling is called *fill factor*. The coefficient of coupling is increased by placing the coil close to the conductor. A higher value of *k*, called "tight" coupling, allows better mutual induction. "Loose" coupling, with a low value of *k*, has the opposite effect.

Chapter 6
Inductive Reactance and Impedance

Inductive Reactance

When alternating current (AC) is an inductance, there is less current than the resistance alone would allow. This additional opposition to sine-wave alternating current in an inductance is its *inductive reactance* X_L or ωL. The counter emf produced by inductance is its reaction against a change of current.

To calculate the amount of inductive reactance opposing a sine-wave alternating current, the following formula can be used:

$$X_L = 2\pi f L \text{ or } X_L = \omega L \qquad \text{(Eq. 6-1)}$$

where

X_L = inductive reactance in ohms,
f = frequency in Hz,
L = inductance in henrys, and
ω = $2\pi f$.

Inductance L depends on the construction of the coil. Inductive reactance X_L depends on the frequency f of the alternating current.

When direct current (DC) is applied to a coil, only resistance R opposes the current flow. However, when alternating current is applied to the same coil, the changing magnetic field also induces a current in the coil that opposes the original current. This opposing current is a result of the magnetic field cutting across windings in the coil. This additional opposition to current flow in an AC circuit is called *self-induction.*

For direct current, the inductive reactance X_L is zero and only resistance R opposes the current flow, as shown in Fig. 6-1(a). However, in an AC circuit, both resistance R and inductive reactance X_L oppose the current flow, as illustrated in Fig. 6-1(b).

Inductive reactance increases with inductance because a higher value of L can produce more induced voltage to oppose the current. Similarly, a higher frequency allows more induced voltage and a higher inductive reactance. The constant 2π (in Eq. 6-1) is derived from the circular motion related to a sine-wave cycle. The inductive reactance increases in proportion to inductance and frequency. When f is doubled, X_L is doubled. This linear increase of X_L with frequency is illustrated in Fig. 6-2. Similarly, Fig. 6-3 shows that X_L increases linearly with higher values of inductance L.

Inductive Circuits

In addition to the fact that X_L limits the amount of alternating current, the back emf induced across the inductance makes the instantaneous values of current in the inductance lag in time behind the

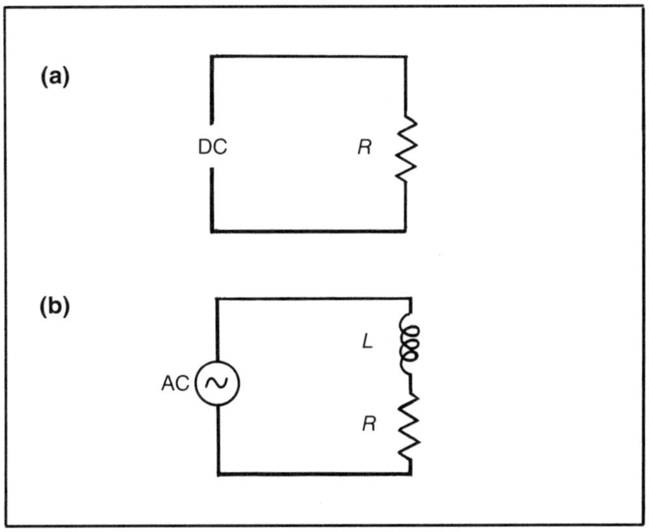

Figure 6-1. Opposition to current flow. (a) Resistance in a DC circuit. (b) Resistance and inductance in an AC circuit.

instantaneous voltage values. The frequency is the same for the current I in the inductance and for the voltage E across it, but they are out of phase. This phase shift results because the voltage induced across the coil opposes any change in current through the inductance. With sine-wave current, the phase angle is 90 degrees between the lagging inductive current and its induced voltage. The resistance does not shift the phase of the current. When a circuit has R and X_L, the phase angle is between 0 degrees and 90 degrees.

With sine-wave variations of current producing an induced voltage, the current lags its induced voltage by exactly 90 degrees, as shown in Fig. 6-4. The inductive circuit in (a) has the current and voltage waveshapes shown in (b). The vectors in (c) show the 90 degree phase angle between I and E_L.

The self-induced voltage E_L value depends on the rate of change of current because it corresponds to motion of the magnetic flux as it expands and collapses. The polarity of the induced voltage depends on the direction of the change in current with its associated field. A field moving outward as it expands produces one polarity of induced voltage; the same field collapsing into the inductor produces the opposite voltage polarity.

Details of the varying current I_L and its induced voltage E_L are shown in Fig. 6-5. Consider that this is not the first cycle, but that the current is at the

$X_L = 2\pi fL$, ohms	Frequency, Hz
0	0
20	10
40	20
100	50
200	100
400	200
1,000	500
10,000	5,000

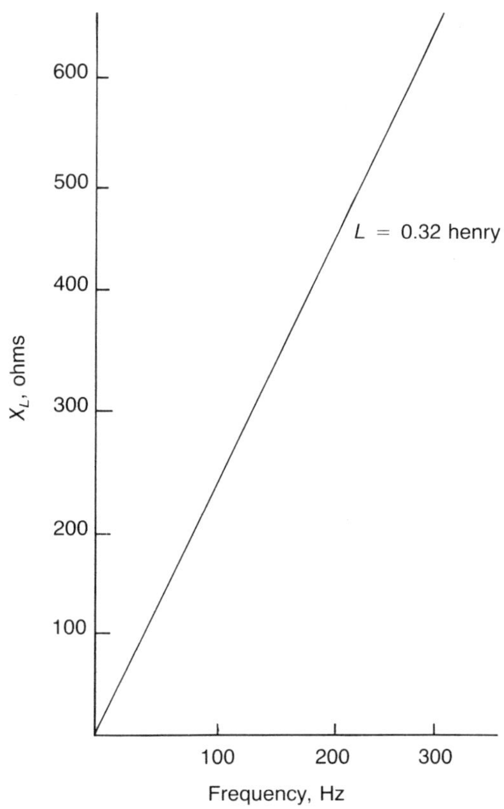

Figure 6-2. Linear increase of X_L with higher frequencies.

$X_L = 2\pi fL$, ohms	L, henrys
0	0
20	0.032
40	0.064
100	0.160
200	0.320
400	0.640
1,000	1.60
10,000	16.0

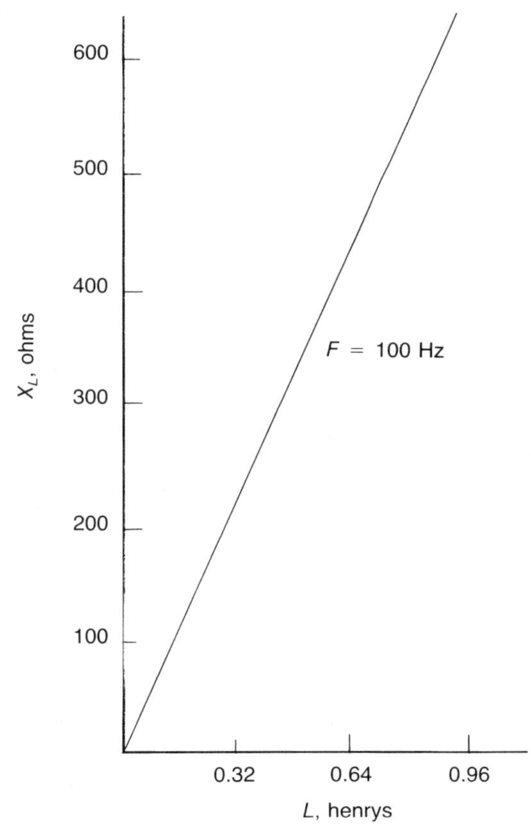

Figure 6-3. Linear increase of X_L with higher values of inductance.

zero point after passing through negative values in the preceding cycle.

At point a, then, the current changes from zero to a positive value. Note that *I* increases sharply here as the waveform has its greatest slope upward. The magnetic field at this time then expands most rapidly, producing maximum induced voltage. At point b, *I* is still increasing, but its field is expanding less rapidly because the current has less slope, corresponding to a slower rate of change. At the 90 degree point at c, it has its maximum value but the waveform is flat at the top, meaning zero slope and no change in current. At this point, therefore, the induced voltage is zero.

Between 90 degrees and 180 degrees, the current is decreasing instead of increasing—the field is collapsing, corresponding to the opposite direction for motion of the flux. The result is an opposite induced voltage of negative polarity. The amount of negative induced voltage increases as the current declines from point c to point d, because the rate of change in the current is greater as the sine wave approaches the horizontal axis. Maximum negative induced voltage occurs at 180 degrees, where the current goes through its zero value, because at this point the rate of change of current is greatest as the field collapses.

From 270 degrees to 360 degrees, the reversed collapsing. This collapse induces the same voltage polarity that the counterclockwise field did when expanding during the first 90 degrees as the directions of both the field lines and the motion are reversed. For the end of the cycle, at 360 degrees, the current goes through its zero value again in the positive direction when the induced voltage repeats its maximum positive value at 0 degrees.

The same action occurs for each succeeding cycle of current. The current and voltage waveforms then are exactly 90 degrees out of phase because the minimum value of one corresponds to the maximum value of the other.

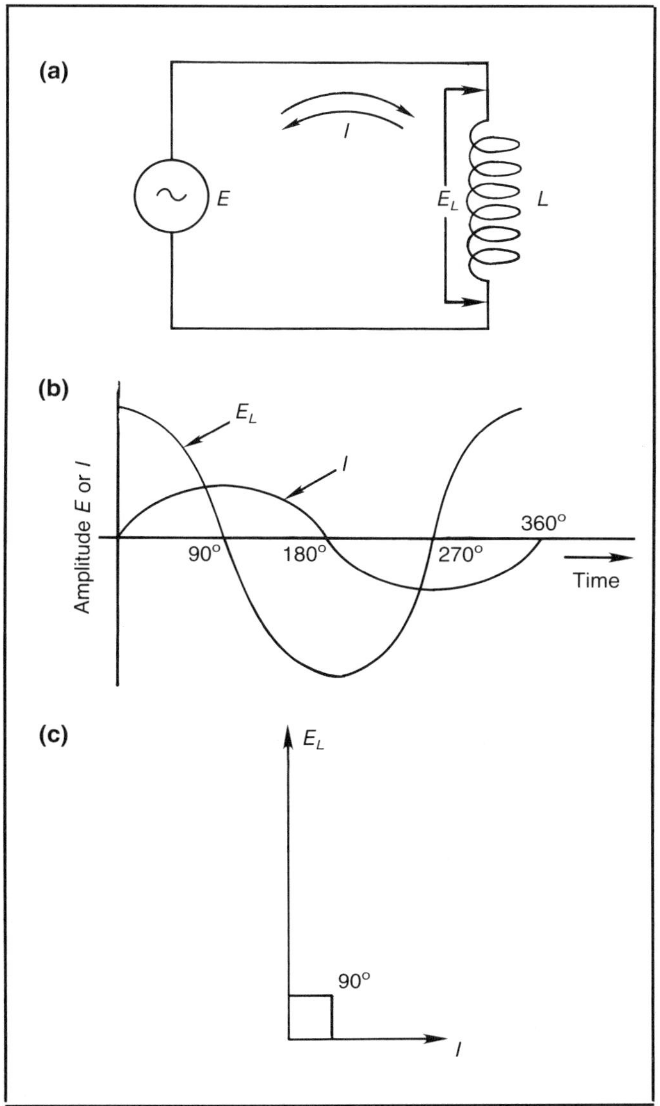

Figure 6-4. Current in an inductance lags 90 degrees in time behind the voltage across the inductance. (a) Circuit. (b) Sine wave of *I* lags E_L by 90 degrees. (c) Vector diagram.

Inductive Reactance and Resistance in Series

When a coil has series resistance, the current is limited by both X_L and R. This current is the same

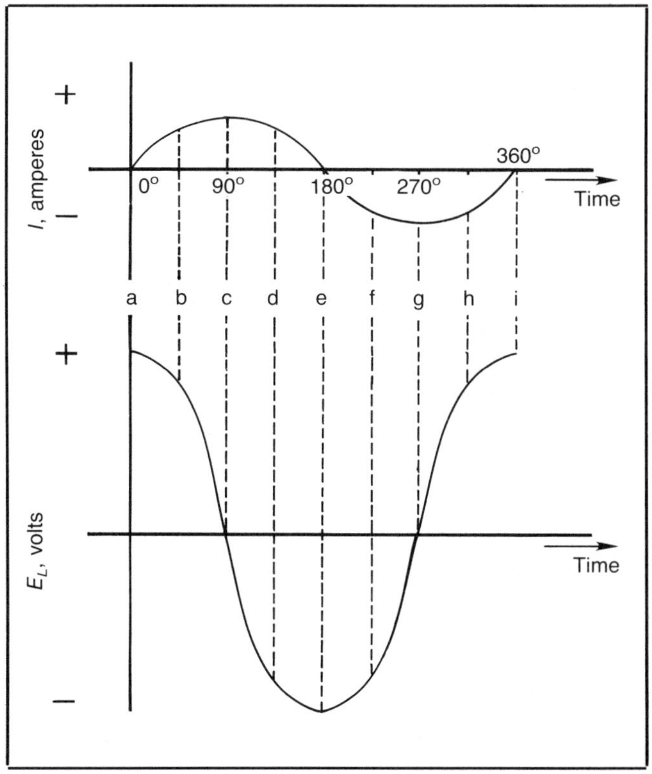

Figure 6-5. Induced voltage E_L is maximum when *I* is zero and E_L is zero when *I* is maximum, resulting in 90 degree phase angle between E_L and *I*.

in X_L and R because they are in series, as shown in Fig. 6-1(b). Each has its own voltage drop, equal to IR for the resistance and IX_L for the reactance. If the inductance is considered alone, its voltage drop E_L still leads the series current I by 90 degrees; that is, I lags E_L by 90 degrees. The IR voltage has the same phase as the current; however, because the resistance has no reactance, E_R lags E_L by 90 degrees. This lag is shown by the phase relation in Fig. 6-6. The current is the phase reference for series circuits because I is the same in X_L and R. The resistive voltage IR has the same phase as the current. Both IR and I are 90 degrees lagging E_L, however, because the current in an inductance lags the voltage across it by 90 degrees.

When the E_R voltage wave is combined with the E_L voltage wave, the result is the voltage wave for the applied generator voltage E. The sum of E_R and E_L equals E because the sum of series voltage drops must add to equal the applied voltage.

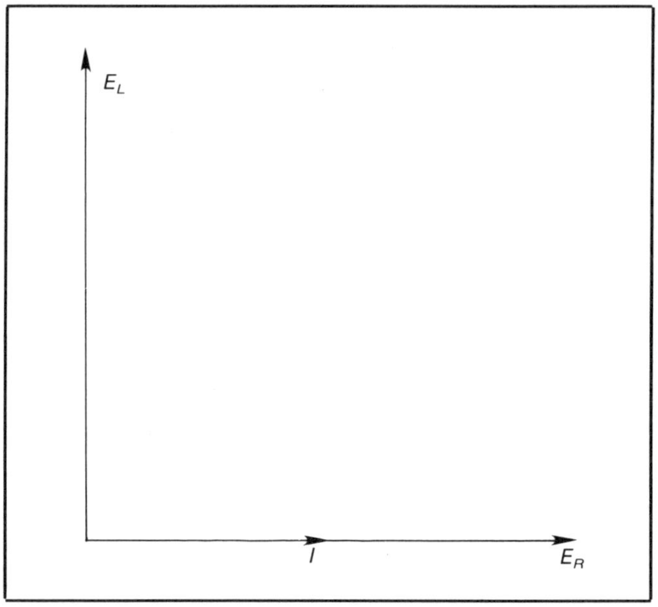

Figure 6-6. Vector diagram of phase relations.

Vector Voltage Triangle

Instead of combining waveforms that are out of phase, they can be added more quickly by using their equivalent vectors, as shown in Fig. 6-7.

The method is to add the tail of one vector to the arrowhead of the other, using the angle required to show their relative phase. E_R and E_L are at right angles because they are 90 degrees out of phase. The sum of the vectors is a resultant vector from the start of one to the end of the other. Because the E_R and E_L vectors form a right angle, the resultant vector is the hypotenuse of the right triangle.

From the geometry of a right triangle, the Pythagorean theorem states that the hypotenuse is equal to the square root of the sum of the squares of the sides. For the voltage triangle in Fig. 6-7(a), the resultant vector voltage is

$$E = \sqrt{(E_R^2 + E_L^2)} \qquad \text{(Eq. 6-2)}$$

where E is the vector sum of the two voltages E_R and E_L 90 degrees out of phase. All values of E must be in the same units.

Impedance Triangle

A vector triangle of R and X_L in series corresponds to the voltage triangle, as shown in Fig. 6-7(b), but the common factor I cancels because the current is the same in X_L and R. The resultant of the vector addition of R and X_L is their opposition in ohms. The combined opposition of resistance and reactance in an AC circuit is called impedance. Its symbol is Z, and the unit is ohms. For the impedance triangle of a series circuit with reactance and resistance, therefore,

$$Z = \sqrt{(R^2 + X_L^2)} \qquad \text{(Eq. 6-3)}$$

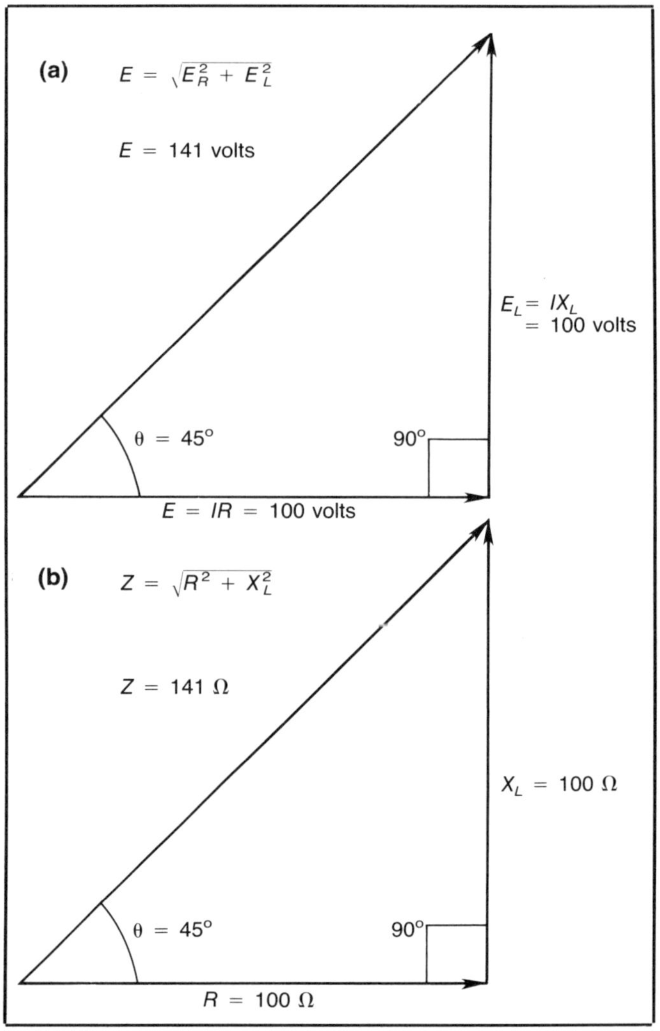

Figure 6-7. Vector addition of two components 90 degrees out of phase. (a) Triangle of series resistive and reactive voltages. (b) Similar triangle of series resistance and reactance.

with R and X_L in ohms, Z is also in ohms. For example, in Fig. 6-7(b),

$$Z = \sqrt{(100)^2 + (100)^2} = \sqrt{(20{,}000)} = 141 \text{ ohms}$$

Phase Angle with Series X_L

The angle between the generator voltage and its current is the *phase angle* of the circuit. Its symbol is θ (theta). In Fig. 6-7(a), the phase angle between E and I is 45 degrees. In the corresponding impedance triangle in Fig. 6-7(b), the angle between Z and R is also equal to the phase angle.

Therefore the phase angle can be calculated from the impedance triangle of a series circuit as follows:

$$\cos\theta = \frac{R}{Z} \qquad \text{(Eq. 6-4)}$$

The cosine (cos) is a trigonometric function of an angle, equal to the ratio of the adjacent side to the hypotenuse in a right triangle. In this impedance triangle, the resistance R is the side adjacent to the phase angle and the impedance Z is the hypotenuse. In Fig. 6-7(b), for example,

$$\cos\theta = \frac{R}{Z} = \frac{100}{141} = 0.707$$

Because a trigonometric function is a ratio, it has no units. The angle that has a cosine value of 0.707 is 45 degrees. Therefore, the phase angle for this example is 45 degrees.

The basic trigonometric functions for a right triangle are illustrated in Fig. 6-8. Sometimes it is more convenient to calculate the phase angle from the resistance and reactance, without figuring the impedance. This calculation can be made in terms of the tangent (the ratio of the opposite side to the adjacent side) of any angle because the side opposite the angle θ in the triangle of Fig. 6-7(b) is the reactance X_L and the adjacent side is the resistance R,

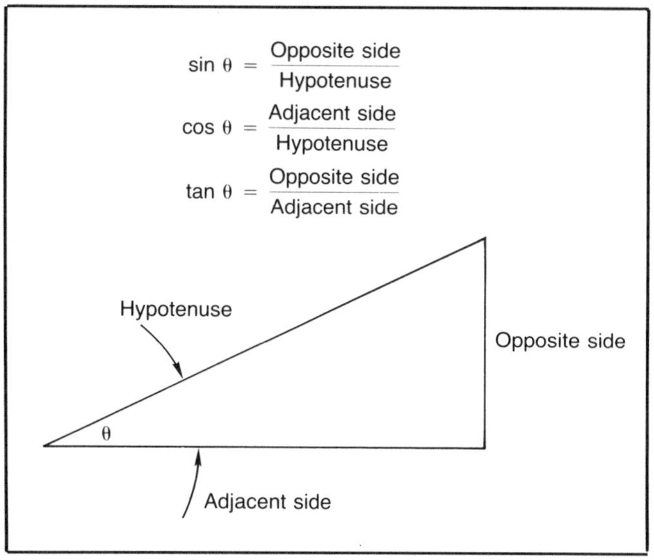

Figure 6-8. Trigonometric functions for a right triangle.

$$\tan\theta = \frac{X_L}{R} \qquad \text{(Eq. 6-5)}$$

Calculating the phase angle in terms of the tangent

$$\tan\theta = \frac{X_L}{R} = 1 \qquad \text{(Eq. 6-6)}$$

from a trigonometric table, the angle that has the tangent equal to 1 is 45 degrees. Therefore the phase angle in this example is 45 degrees, the same answer as with the cosine formula. Either can be used: $\cos\theta$ is more convenient if Z is known; $\tan\theta$ is easier to calculate from the reactance and resistance without knowing Z.

In a series circuit, the higher the value of X_L compared with R, the more inductive the circuit is. This means there is more voltage drop across the inductive reactance and the phase angle increases toward 90 degrees.

Chapter 7
Eddy Current Test Principles

Basic Principles

In many eddy current or electromagnetic induction tests, the test object is placed in the varying magnetic field of a coil carrying an alternating current (AC). The AC magnetic field induces eddy currents in the test object. These eddy currents, in turn, produce an additional AC magnetic field in the vicinity of the test object. Fig. 5-3(a) illustrated conditions for a coil placed on the surface of a test object. The vector H_0 represented the primary AC field of the test coil, whereas H_r indicated the reaction field resulting from eddy currents in the test object. Fig. 5.3(b) represented conditions for the arrangement in which the test coil surrounds the test object. In each case, two AC magnetic fields are superimposed. The magnetic field near the test coil is modified if a test object is present.

Test-Coil Characteristics

In general, the test coil is characterized by two electrical impedance quantities:

1. The inductive reactance $X_L = 2\pi fL$, where f is the frequency of the AC field in cycles per second (Hz) and L is the self-inductance of the coil.
2. The ohmic resistance R.

It is common practice to plot the reactance X_L as ordinate and resistance R as abscissa in the impedance plane. In this way, the test-coil impedance Z is represented by a point P, formed by two perpendicular ("normal") components, $X_L = 2\pi fL = \omega L$, and R, on the impedance plane. In the absence of a test object, the empty test coil has a characteristic impedance, with coordinates $X_{L_0} = \omega L_0$, and R_0, shown on the impedance plane by the empty point ("air point") P_0 of Fig. 7-1. If the test object is placed in the field of the test coil, the original field of the empty test coil is modified by the superimposed field of the eddy currents. This field modification has exactly the same effect as would be obtained if the characteristics of the test coil had been changed. The influence of the test object can be described by a variation in test-coil characteristics. The apparent impedance of the "empty" coil represented by P_0 is displaced to P_1 (corresponding to new values of reactance L_1 and resistance R_1) under the influence of the test object (see Fig. 7-1).

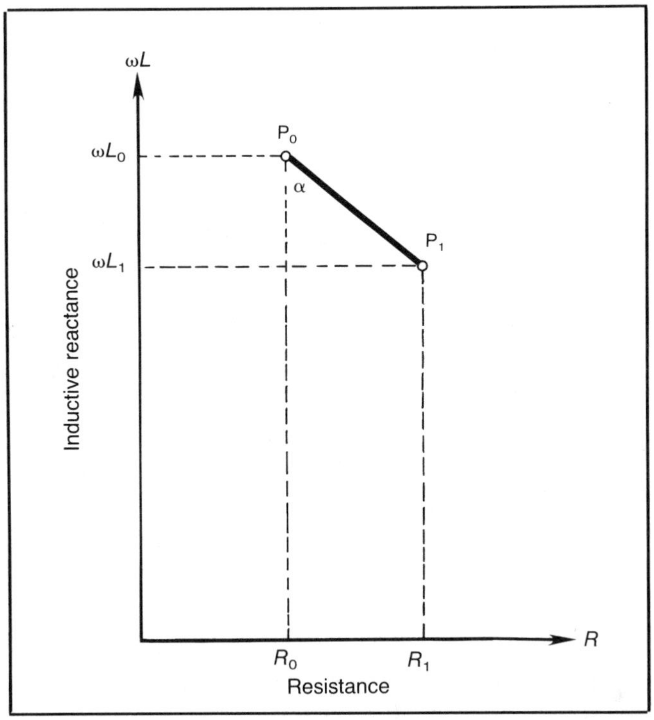

Figure 7-1. Representation of test-coil characteristics on impedance plane. Point P_0 = impedance of coil in absence of test object. Point P_1 = impedance of test coil with test object.

The magnitude and direction of displacement of the apparent impedance from P_0 to P_1 under the influence of the test object are functions of the properties of the test object and the characteristics of the instrumentation. Significant properties of the test object include

1. electrical conductivity (σ) or resistivity (ρ),
2. dimensions (thickness, diameter, etc.),
3. magnetic permeability (μ), and
4. presence of discontinuities, such as cracks or cavities.

Significant instrument characteristics include

1. frequency of the AC field of the test coil,
2. size and shape of the test coil, and
3. distance of the test coil from the test object (coupling between the coil and the specimen).

The influence of several physical properties of the test specimen upon the impedance characteristics of the probe coil can be calculated for various test frequencies in certain cases. Often it is possible to determine from the impedance change (distance P_0 to P_1), and frequently to measure, quantitatively and independently of each other, not only the conductivity, dimensions, and magnetic permeability of the test object but even the magnitude and direction of cracks. The optimum frequency for a specific test problem is determined by theory or by experiment and provides the highest sensitivity obtainable with eddy current methods for detection of

1. variations in electrical conductivity or permeability, used in sorting alloys,
2. variations in test-object diameter or thickness, and
3. cracks.

Indications are obtained without electrical contact with the test object, in extremely short time intervals. Physical properties measured include alloy, heat treatment, hardness, depth of case, surface decarburization, magnitude of defects, and dimensions.

Eddy Current Probe-Coil Test Arrangement

Fig. 5-3 shows typical arrangements of eddy current test probe coils and conducting test materials. Fig. 5-3(a) shows an eddy current probe coil adjacent to the flat surface of an electrically conducting test-material sheet. The probe-coil winding consists of several turns of wire wrapped around a circular coil; the turns lie parallel to the surface of the material under test. The axis (centerline) of the probe coil is perpendicular to the test-material surface. The axial component of the probe coil's magnetic field intercepts or couples with the test material, at least with the near-surface layers of the test material. When AC flows in the probe-coil winding, its magnetic field induces eddy current in the sheet material. The eddy current reaction field H_r acts along the direction of the probe-coil axis so as to oppose the magnetizing-coil field H_0. The total or net magnetic-field intensity H_e is the phasor (vector) sum of the magnetizing-coil field H_0 and the eddy current reaction field H_r, as shown in Eq. 7-1:

$$H_n = H_0 + H_r^* \qquad \text{(Eq. 7-1)}$$

Eddy Current Encircling-Coil Test Arrangement

Fig. 5-3(b) shows an encircling-coil eddy current test arrangement. The encircling-coil winding consists of several turns of wire wrapped on a hollow, cylindrical coil form. Cylindrical test objects can be placed within the coil or continuously passed through it during tests of long rods or tubes. The axis of the encircling coil is parallel to and typically concentric with the longitudinal axis (or centerline) of the round bar or tube to be tested. Within the encircling coil, the magnetic field lies parallel to the coil's axis (in the same direction as the length of the long bar or tube). This produces longitudinal magnetization within the conducting test object. The circular paths of the induced eddy currents are parallel to the cylindrical outer surface of the test material. These eddy current paths encircle the longitudinal magnetic flux lines in the test material.

Also shown in Fig. 5-3(b) is the direction of the magnetizing-coil field H_0 (at a particular instant in its alternating field). As always, the induced eddy currents tend to oppose the applied magnetic field that induced them. The direction of the eddy current reaction field H_r is shown as longitudinal within the cylindrical bar or tube, in the direction opposing the applied field H_0. The direction, in space, of the net magnetic field H_n is also parallel to the length of the test material, at the center of the encircling magnetizing coil.

Transformer Action of Eddy Current Magnetizing Coils

The basic eddy current test magnetizing system is analogous to a simple two-winding electrical transformer (see Fig. 5-2). The magnetizing coil serves as the primary winding of this transformer. The conducting material near the excited surface of the test material serves as the one-turn, short-circuited secondary winding of this transformer. The resistance of the eddy current path within the

* Vector quantities are often represented by a bar or arrow over the quantity, i.e., $\overline{H}$ or $\vec{H}$. They may also appear in bold type.

test material serves as a purely resistive load for the transformer. The eddy currents in the test material are the secondary current.

Transformer Impedance with Empty Test Coil

In the absence of the test material, known as the *empty-coil condition*, the transformer operates in the no-load condition. The primary (or magnetizing winding) current is then determined by the impedance of the primary winding alone. This empty-coil impedance Z_0 consists of a resistive component R_0 and an inductive reactance component, X_{L_0}. The total empty-coil impedance is given by Eq. 7-2.

$$Z_0 = R_0 + jX_{L_0} \qquad \text{(Eq. 7-2)}$$

where j is an imaginary term that causes a linear increase in phase-angle lag with depth.

Inductive reactance is given by Eq. 7-3:

$$X_{L_0} = 2\pi f L_0 = \omega L_0 \qquad \text{(Eq. 7-3)}$$

Significance of the Empty Test-Coil Impedance

The empty test coil impedance, given by Eq. 7-2, is a useful reference characteristic of the eddy current test coil system; all other test signals can be compared to it. With the empty test coil, the winding resistance R_0 is typically very small in comparison to the magnitude of its inductive reactance. The magnitude of this impedance is given by Eq. 7-4:

$$Z_0 = \sqrt{(R_0^2 + X_{L_0}^2)} \qquad \text{(Eq. 7-4)}$$

If the test-coil resistance is negligible, the empty test coil impedance can be assumed to equal its inductive reactance, as in the approximation of Eq. 7-5:

$$Z_0 \cong X_{L_0} \cong \omega L_0 \qquad \text{(Eq. 7-5)}$$

As will be discussed later, it is possible to *normalize* eddy current test signal magnitudes by dividing all test signals by the empty-coil test signal magnitudes. Normalization of eddy current test signals offers a remarkable advantage: tests with different sizes of test coils and differing types of test materials can be related to universal curves on the complex signal plane; the curves are then applicable to numerous test conditions.

Transformer Loading Effect of Eddy Currents in Nonmagnetic Test Materials

When a nonmagnetic conducting test material is brought into the magnetic field of an eddy current magnetizing coil, it has two immediate effects upon the transformer characteristics of the coil: (1) the eddy currents represent a resistive load in the secondary circuit of the magnetizing-coil transformer system, and (2) the eddy current reaction field decreases the magnetic flux linkages with the magnetizing coil and so lowers the inductive reactance of the magnetizing coil.

The general result of introducing a nonmagnetic conducting test material is an increase in the magnitude of the apparent resistance of the magnetizing coil and a decrease in the magnitude of the apparent inductive reactance of the magnetizing coil. This results in a change in the position of the test signal point on the complex-plane diagram for the eddy current test signals (see Fig. 7-1).

Complex-Plane Diagrams for Test Signals with Nonmagnetic Test Materials

Fig. 7-2 shows examples of eddy current test signal points on the complex-plane diagrams for the cases of (a) the magnetic field intensity, (b) the magnetizing-coil impedance, (c) the magnetizing-coil (or coincident-voltage pickup coil) voltage signal, and (d) the effective permeability. These complex-plane diagrams for the various types of eddy current test signals are essentially equivalent when the effects of nonmagnetic test materials upon the displacement of the test-signal points are considered.

In each diagram of Fig. 7-2, the horizontal component of the test-signal point displacement corresponds to the resistive (or real) component of the eddy current test signal. In the case of the magnetizing-coil impedance plane only, this signal component includes the actual resistance of the magnetizing-coil winding and the additional effective resistance caused by eddy current losses within the test material. In the other complex-plane diagrams of Fig. 7-2, the magnetizing-coil resistance is excluded from the resistive component of the test-signal point displacement in the horizontal direction of the real signal component.

In each complex-plane diagram of Fig. 7-2, the vertical (imaginary) component of the test-signal point displacement corresponds to the effect of the magnetic flux linkages of the net magnetic field flux. The eddy currents flowing within the nonmagnetic test material always act to oppose the magnetizing-coil field and so tend to lower the magnitude of the flux linkages with the magnetizing coil or with a separate voltage-signal pickup coil. The same effect is observed when the test signal is derived from a magnetic field intensity detector, such as a semiconductor magnetic field detector or a small voltage pickup coil.

Eddy Current Test-Signal Variations with Nonmagnetic Test Materials

The empty-coil test signal point with no conducting material located within the magnetizing-coil field is shown by point P_0 in each of the complex-plane diagrams of Fig. 7-2. This empty-coil

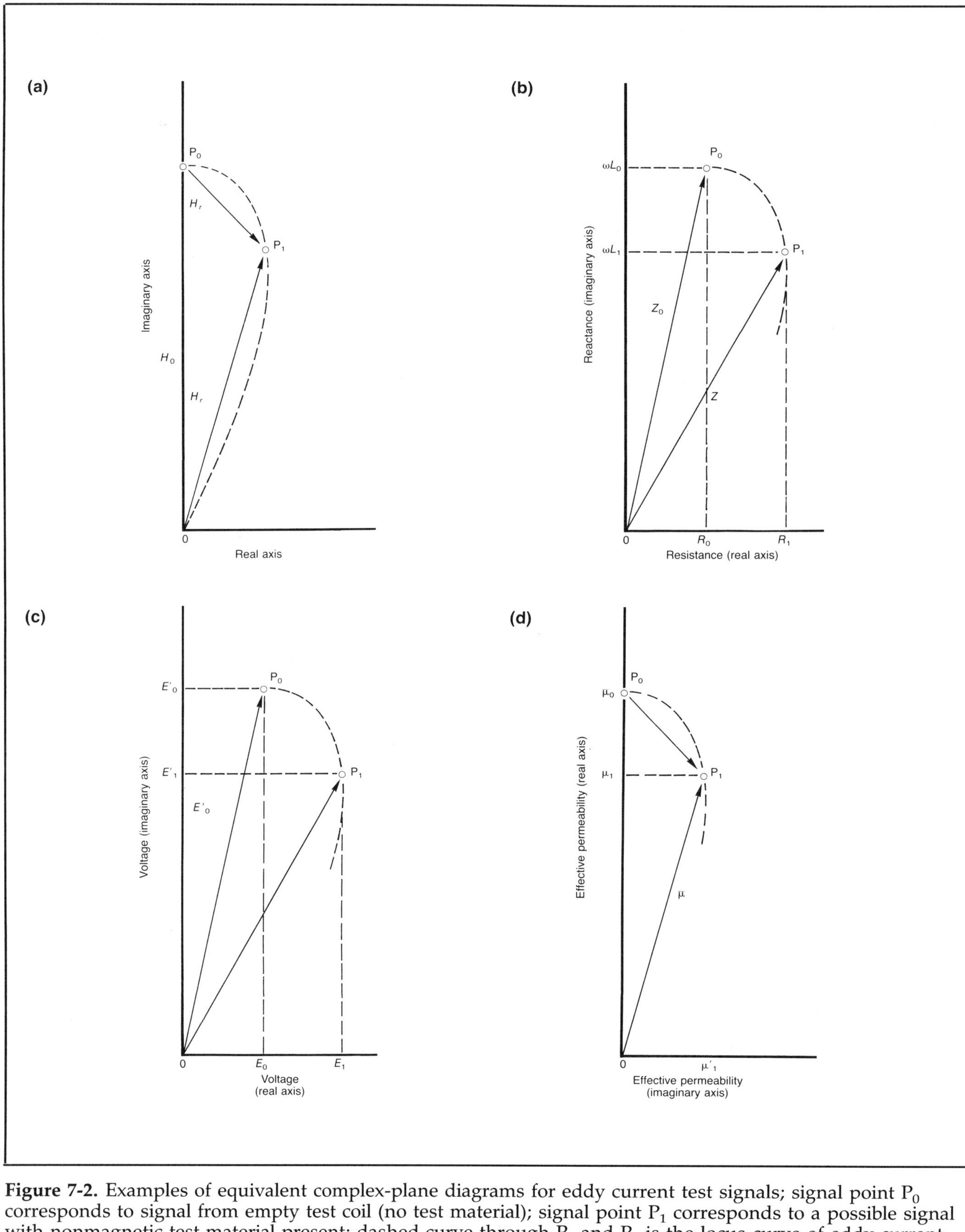

Figure 7-2. Examples of equivalent complex-plane diagrams for eddy current test signals; signal point P_0 corresponds to signal from empty test coil (no test material); signal point P_1 corresponds to a possible signal with nonmagnetic test material present; dashed curve through P_0 and P_1 is the locus curve of eddy current test signals. (a) "Ideal" complex-plane diagram, referring only to magnetic field strength vectors ($\overline{H}_0$, $\overline{H}_r$). (b) Complex impedance-plane diagram ($X_L = \omega L$, and Z is impedance). (c) Complex-plane diagram referring only to voltages. (d) Complex-plane diagram referring only to permeability μ (note switching of real, imaginary axes).

signal point is always represented by the upper left extremity of the test-signal locus curves on the complex-plane diagram. Dashed lines show these locus curves in Fig. 7-2.

Because this point is well-known by most workers in the field of eddy current testing, the empty-coil signal point ("air point") is not always labeled in complex-plane diagrams for eddy current tests. The empty-coil signal point location is a useful reference to which all other eddy current test signals (resulting from test materials) can be readily compared.

Elimination of Primary Magnetizing-Coil Resistance from Test Signals

The actual resistance R_0 of the magnetizing-coil winding has no significance in interpretation of eddy current test signals. The important factor is the resistive effect of the test material as reflected into the primary winding. Thus, in many types of eddy current test instrumentation, provision is made to subtract the actual primary-winding resistance from the total apparent resistance to indicate only the resistive effect caused by the test object. The effective resistance is given by Eq. 7-6:

$$\text{effective resistance} = R - R_0 \qquad \text{(Eq. 7-6)}$$

Factors Affecting Resistive Component of Test Signals

The resistive (or real) component of eddy current test signals is highly important in eddy current tests of nonmagnetic test materials. It is a measure of the ohmic resistance losses that occur within the test material during eddy current tests. The energy losses (from the interaction of the eddy currents and the resistance) result in heating of the test material. Their magnitude increases as (1) test frequency is increased, (2) the test magnetizing field strength is increased, and (3) the test material electrical conductivity or thickness (within limits) is increased. The composition of many nonmagnetic alloys is indicated by differences in their electrical conductivities. Eddy current tests for sorting nonmagnetic test materials by alloy or dimensions usually depend primarily upon the variations in the resistive component of the impedance. Some test systems use the loss-sensing principle in their indications of eddy current test signals.

Effects of Temperature and Thermal Drift in Nonmagnetic Test Materials

With metals and alloys, the test material conductivity depends upon the material temperature. This permits noncontacting eddy current techniques of inductive thermometry, as developed by Hugo Libby. However, with the test materials having high electrical conductivity, it is possible for the heat produced by the eddy current ohmic resistance losses to raise the temperature of the test materials, especially in tests made with higher test frequencies or with higher magnetizing-field strengths. This heating changes their electrical conductivities, leading to an apparent drift in eddy current test indications. This is not a fault in the eddy current test instrumentation, but rather is a true reflection of the change in test-material conductivity induced by a temperature rise. In general, eddy current tests should be made on test materials in equilibrium, with ambient temperatures, and in an environment with nearly constant temperatures to avoid the need for temperature correction.

The problem of slowly varying test-material temperatures can be avoided during defect detection or sorting of materials by the use of differential or comparison test systems, where the unknown test object is compared with a similar test object, or with another portion of the same test object, subject to the same temperature conditions.

Magnetizing-Coil Inductive Reactance Characteristics

In addition to its response to the resistive components of the eddy current test signal, the eddy current magnetizing coil responds to the magnetic flux that passes within its winding. The inductance L of the magnetizing (primary) coil winding is given by Eq. 7-7:

$$L = \frac{N_p\phi}{I_p} \qquad \text{(Eq. 7-7)}$$

The magnetic flux linkages with the primary magnetizing coil, λ, are given by Eq. 7-8:

$$\lambda = N_p\phi \qquad \text{(Eq. 7-8)}$$

The inductive reactance X_L of the primary magnetizing coil is given by Eq. 7-9:

$$X_L = \omega L = 2\pi f L \qquad \text{(Eq. 7-9)}$$

For Eqs. 7-7 to 7-9,

N_p = number of series winding turns in magnetizing coil,
I_p = primary winding current (amperes),
L = primary magnetizing coil inductance (henrys),
λ = flux linkages with magnetizing coil (weber-turns),
ϕ = magnetic flux encircled by primary turns (webers),
X_L = primary magnetizing coil inductive reactance (ohms), and
ω = $2\pi f$ = angular frequency (radians/second).

Inductance and Inductive Reactance of Empty Test Coil

The inductance L and the inductive reactance X_L of the empty test coil (in the absence of conducting test material) are the same as for the test coil in air. In this case, the magnetic flux, which couples with the coil winding turns, is generated by the magnetizing field, which is in turn created by the AC flowing in the coil winding. Mathematical solutions exist for the magnetic field of plane circular test coils (probe-coil fields) and for the case of the infinitely long helical coil (encircling test coil). Alternatively, the inductance or the inductive reactance of the empty test coil can be determined by use of AC bridge circuits and standard techniques for measurement of these characteristics. In many cases, the test-coil designer uses this type of measurement to evaluate various designs of eddy current test coils.

Because the inductance and the inductive reactance (at any specific test frequency) of the empty test coil remain fixed and constant, they provide a stable reference base to which all other eddy current test conditions can be compared. In addition, when two test coils are to be made identical (in these characteristics) for the purpose of eddy current comparison tests, AC bridge measurements of inductance or of inductive reactance can be used to prove that their characteristics are equivalent, within acceptable tolerance limits. The easy way to adjust test coils that are not identical is to add or subtract winding turns (or parts of turns) until the empty-coil resistances and inductances become equivalent. In this process, it is important to recognize that, for the empty test coil, the coil inductance is proportional to the square of the number of series winding turns. On the other hand, the coil resistance R_0 is directly proportional to the number of winding turns when the coil diameter is held constant.

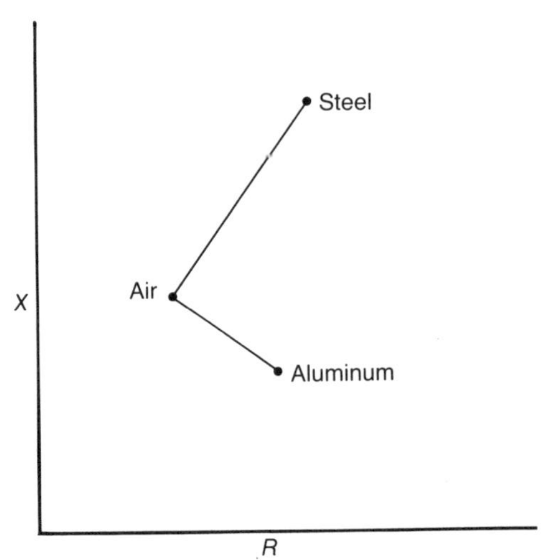

Figure 7-3. Impedance-plane CRT presentation showing changes in inductive reactance and resistance with magnetic and nonmagnetic materials.

Test-Coil Reactance with Nonmagnetic Materials

When a conductive nonmagnetic test material is placed in the field of an eddy current test magnetizing coil, the induced eddy currents provide a magnetic reaction field that opposes and weakens the total (or net) magnetic flux density. This effect results from the opposing ampere-turns provided by the eddy currents in the test material. As the magnetic flux density decreases, it reduces the magnetic flux linkages with the turns of the magnetizing coil. This lowers both the inductance and the inductive reactance of the magnetizing coil. The magnitude of this reduction depends upon (1) the test material conductivity and thickness, (2) the test frequency, and (3) the proximity of the magnetizing-coil winding to the test material.

The highest available magnetic flux density occurs with the empty test coil. When a conducting, nonmagnetic test material is placed within the coil field, the degree of magnetic flux reduction is greater as each of the three factors (listed above) is increased. These field-weakening effects are clearly evidenced with thin layers of nonmagnetic test materials. With very thick layers, exceeding the standard depth of eddy current penetration by a considerable margin, these effects are complicated by the eddy current attenuation and skin effect.

Test-Coil Reactance with Ferromagnetic Test Materials

When ferromagnetic test materials (such as iron or steel) are placed in the magnetizing-coil field, the magnetizing-coil reactance changes in a far different way than with nonmagnetic test materials. As the highly permeable test material is brought into the field of the magnetizing coil, those flux lines that enter the ferromagnetic test material find portions of their path length in a material with far less magnetic reluctance than air. The portion of the path of each such flux line in air is shortened. The magnetizing-coil field then includes increased magnetic flux densities encircled by the coil windings. This increases both the inductance and the inductive reactance of the magnetizing coil (see Fig. 7-3). The maximum values of inductance and of inductive reactance occur when the surface of the magnetic test material is very close to the magnetizing-coil winding. This effect is the opposite of that found with nonmagnetic test materials, where both the coil inductance and its inductive reactance are reduced as the test material is brought close to the magnetizing-coil windings (see Fig. 7-3).

In fact, with high-permeability ferromagnetic test materials, it is possible that nearly half of the flux-line length lies within the ferromagnetic material. The length of the flux path in air would then be reduced to nearly half of that for the empty test coil in air. The ampere-turns of the magnetizing coil are then used primarily to maintain the magnetic flux lines in air. A relatively small portion of these magnetizing ampere-turns is required to maintain the flux within steel, for example. This increases the magnetizing-field intensity (in ampere-turns per unit length) of the air path and so increases the magnetic flux density within the magnetizing-coil windings. The magnetizing-coil inductance and inductive reactance thus increase dramatically when a high-permeability magnetic test material is used.

Techniques for Normalization of Eddy Current Test Signals

The concept of normalized components of eddy current test signals, originated by Dr. Friedrich Förster, is very valuable in eddy current test analyses. By use of normalized test signals, the performance of many different test-coil sizes, at various frequencies and with various types of materials, can be described by a single set of equations and complex-plane diagrams. This provides a universal guide to selection of test conditions and a common method for analysis of eddy current test signals. The procedures for normalization of test signals are given next in this section. Throughout most tests using eddy currents, testing complex-plane diagrams are shown in terms of normalized real and imaginary components of eddy current test signals because of their applicability to many different test situations.

Eddy current test signals are normalized by dividing the effective signal magnitude by the magnitude of the corresponding empty-coil signal. The real (horizontal) and the imaginary (vertical) components of the test signal, as shown on complex-plane diagrams, see Fig. 7-4(a), can be normalized by dividing each of these signal component magnitudes by the magnitude of the empty-coil test signal; see Fig. 7-4(b). A similar technique applies for the case of voltage signals derived from (1) the magnetizing coil voltage, (2) the voltage induced within a voltage signal coil (either coincident with the magnetizing coil or with a small pickup coil within the magnetic field of the coil), or (3) the signal of the magnetic field intensity derived from a semiconductor detector.

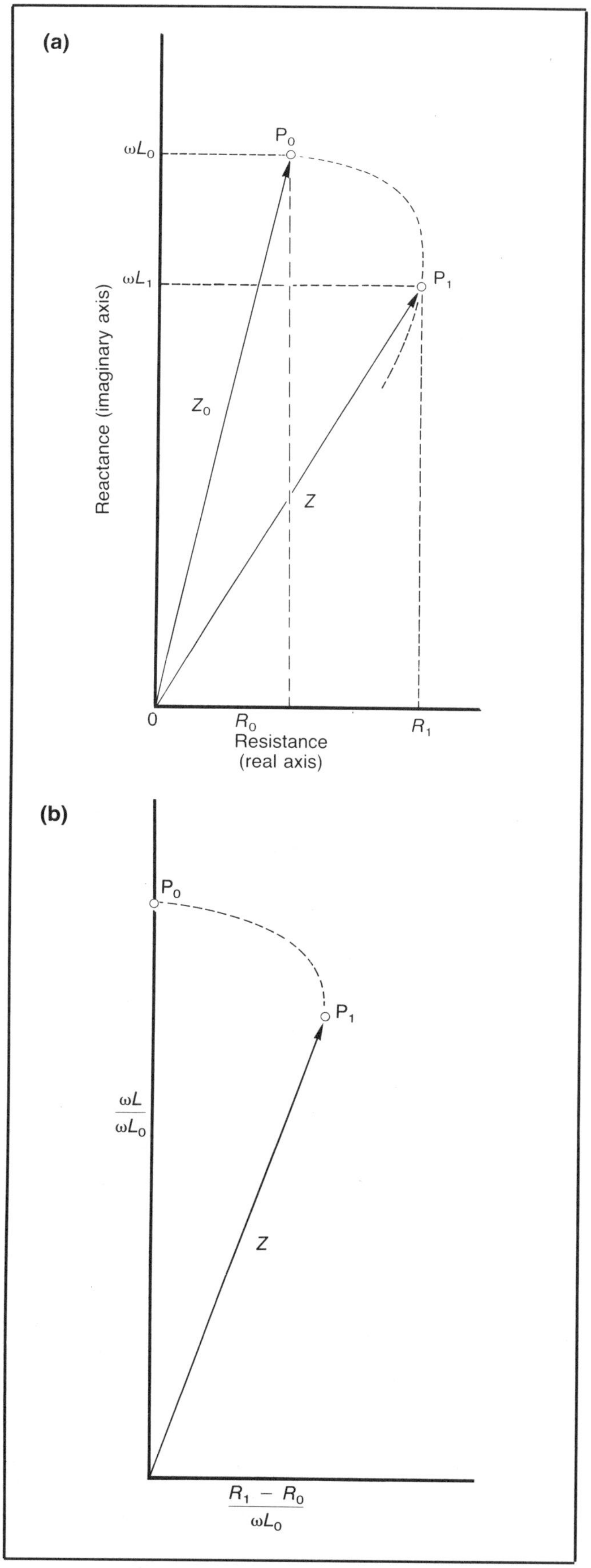

Figure 7-4. Normalization of eddy current test signals. (a) Before normalization. (b) After normalization.

Normalization of Resistive Component of Magnetizing-Coil Impedance

The resistive component of the magnetizing-coil impedance is the real (horizontal) component, as shown on the complex plane (see Fig. 7-4). After elimination of the magnetizing-coil resistance as shown in Eq. 7-6, the effective resistance is divided by the approximation of the empty-coil impedance, as given by Eq. 7-5. The normalized resistive component of the eddy current impedance is then given by Eq. 7-10:

$$\text{normalized resistance} = \frac{(R_1 - R_0)}{\omega L_0} = \frac{(R_1 - R_0)}{X_0} \quad \text{(Eq. 7-10)}$$

Note that in this and succeeding equations for normalized test signals, the normalized signal is a dimensionless ratio. Thus, a normalized eddy current test signal is a valid magnitude, regardless of the specific units of measurement employed.

Normalization of Reactive Component of Magnetizing-Coil Impedance

The inductive reactance component of the magnetizing-coil impedance signal is indicated by the vertical position of the test signal point on the complex plane of coil impedance (see Fig. 7-4). This component is also known as the imaginary component of the magnetizing coil's impedance. The reactive component of the actual test signal is given by Eq. 7-9. To normalize this reactance signal, the actual test signal is divided by the impedance of the empty test coil, as approximated by Eq. 7-5. The normalized inductive reactance component of the test magnetizing-coil impedance is then given by Eq. 7-11:

$$\text{normalized inductive reactance} = \frac{\omega L_1}{\omega L_0} = \frac{X_L}{X_0} \qquad \text{(Eq. 7-11)}$$

In Eq. 7-10 and Eq. 7-11,

ω = $2\pi f$ = angular frequency in radians/s,
L = inductance in henrys,
L_0 = inductance of coil in air in henrys,
L_1 = inductance of coil on test article in henrys,
R_0 = coil resistance in air in ohms,
R_1 = resistive effect of test material in ohms,
X_0 = coil reactance in air in ohms, and
X_L = coil reactance of test article in ohms.

Magnitudes of Normalized Signals with Nonmagnetic Test Materials

When nonmagnetic test materials are placed in the magnetic field of the eddy current test magnetizing coil, the eddy current reaction always tends to reduce the absolute test signal magnitude below that of the empty test coil. Thus, after normalization, the test signal (with test material present) typically has a magnitude not exceeding unity. In fact, the real (resistive) component of the test signal typically does not exceed a value of 0.5 after the signal has been normalized. The imaginary (reactive) component can take on values between 0 and 1.0. The preceding statements apply generally to the case of uniform test materials without discontinuities. When the test material contains discontinuities such as cracks, holes, grooves, and other severe interruptions in the flow path of the eddy currents, local distortions of the magnetic field result and can often be of very high magnitude. Thus, with local detectors such as small coil pickups or semiconductor detectors, discontinuity signals can exceed the value of the empty test coil signal. However, when the voltage pickup coil is of the same size and in the same location as the test magnetizing coil, the normalized test signals with nonferromagnetic test materials usually cannot exceed the magnitude of the empty-coil signal.

Eddy Current Depth of Penetration

The eddy current density (J_X) changes its magnitude with distance from the surface. Its value at a given depth x is given by Eq. 7-12 (after Libby).

$$J_X = J_0 \exp -x\sqrt{(\pi f \mu \sigma)} \qquad \text{(Eq. 7-12)}$$

where

J_0 = current density at surface in A/m^2,
π = 3.1416,
f = frequency in Hz,
μ = magnetic permeability = 4×10^{-7} H/m for nonmagnetic materials,
x = depth from surface in m, and
σ = electrical conductivity in mho/m.

The value of the depth x is called skin depth, or standard depth of penetration, which is represented by the symbol δ. When δ equals depth x, then $J_X/J_0 = e^{-1} = 0.37$, as shown in Fig. 7-5. The standard depth of penetration δ can therefore be defined as the depth at which the current density drops to 37 percent of the current density at the surface. Eq. 7-12 may be written

$$J_X = J_0 \exp\left(\frac{-x}{\delta}\right) \qquad \text{(Eq. 7-13)}$$

where $\delta = (\pi f \mu \sigma)^{-1/2}$ = standard depth of penetration. The word "standard" denotes plane-wave electromagnetic field excitation within the test specimen.

The skin effect causes the currents to be concentrated near the surface adjacent to the excitation coil. The effect increases (i.e., depth decreases) with increasing test operating frequency, test-object electrical conductivity, and test-object

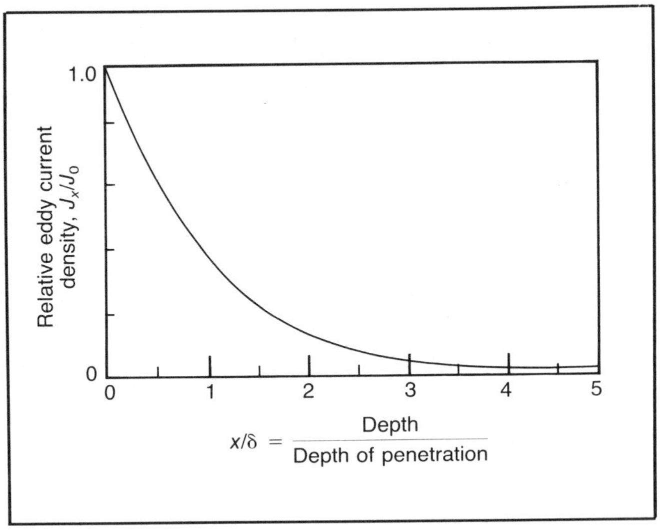

Figure 7-5. Variation of eddy current density as a function of depth, plane-wave case.

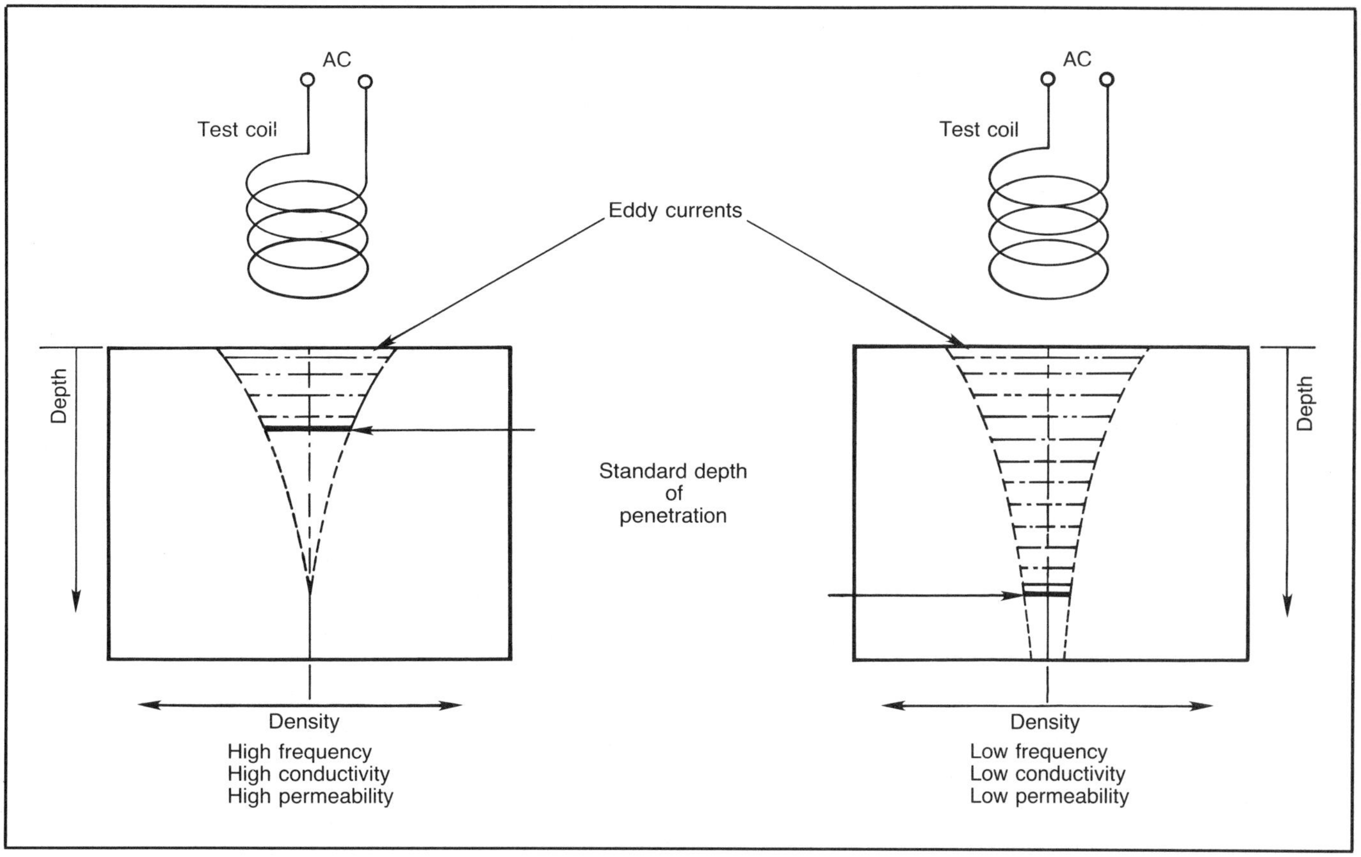

Figure 7-6. Relative effect of frequency, conductivity, and permeability on depth of penetration.

magnetic permeability, as illustrated in Fig. 7-6. The currents decrease exponentially or approximately exponentially with depth, depending upon the test-object shape and thickness (see Fig. 7-5). In addition to the decrease of current density as depth below the surface increases, the phase angle of the current lags increasingly as the depth increases.

Although eddy currents penetrate deeper than one standard depth of penetration, they decrease rapidly with depth. At two standard depths of penetration (2δ), eddy current density has decreased to $(1/e)^2$ or 13.5 percent of the surface density. At three depths of penetration (3δ), the eddy current density has decreased to 5 percent of the surface density.

The phase-angle lag, using the phase angle of the current density at the surface as a reference angle, is given by (after Libby)

$$\theta = \sqrt{x(\pi f \mu \sigma)} = \frac{x}{\delta} \qquad \text{(Eq. 7-14)}$$

where θ = phase-angle lag in radians.

The variation of eddy current phase angle as a function of depth for thick material is shown in Fig. 7-7. When x is equal to one standard depth of penetration, phase lag is 57 degrees, or one radian. The phase lag in degrees is shown in Fig. 7-8 and can be calculated using Eq. 7-15, where x is distance below the surface in the same units, as

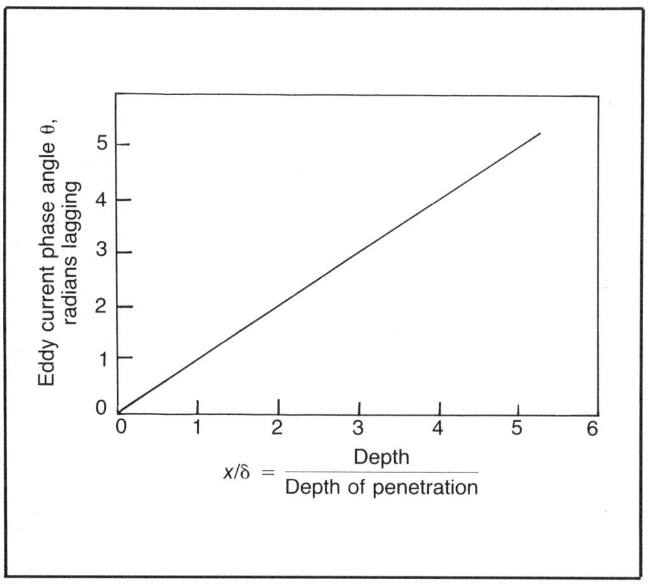

Figure 7-7. Variation of eddy current phase angle as a function of depth, plane-wave case.

$$\theta = 57\left(\frac{x}{\delta}\right) \qquad \text{(Eq. 7-15)}$$

There are various versions of mathematical equations for determining the value of δ, depending on the units and the physical factors used for the calculations. Note, in Table 7-1, $\mu = 1$ for nonmagnetic materials only. To determine the depth in ferromagnetic materials, use the appropriate μ_{rel} value in the equation.

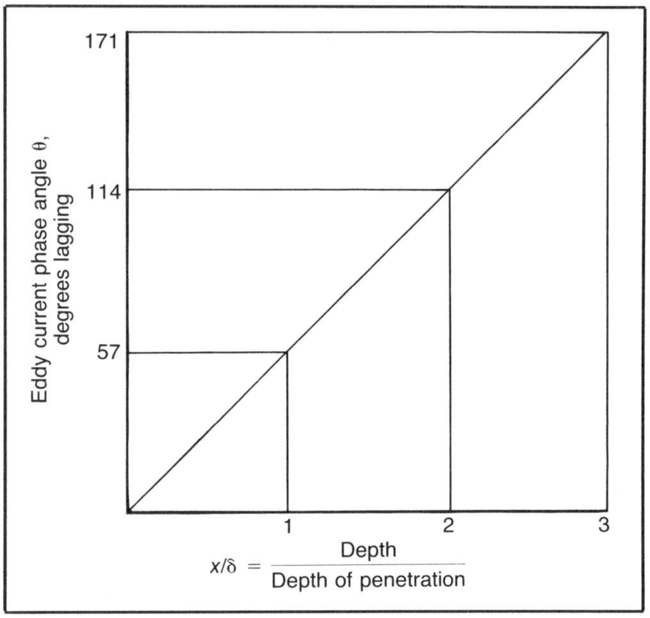

Figure 7-8. Variation of eddy current phase angle in degrees as a function of depth.

One problem in trying to relate the different equations in Table 7-1 is the need for conversion factors.

Fig. 7-9 illustrates the eddy current standard depth of penetration for materials with various conductivities at different frequencies.

Table 7-1
Equations Used for Determining the Standard Depth of Penetration

Hochschild	Libby/ Franklin	Hocking MIL-STD-1537
$\delta = 1/(\pi f \mu \sigma)^{1/2}$	$\delta = (\pi f \mu \sigma)^{-1/2}$	$\delta = 50(\rho/f\mu)^{1/2}$
δ = cm	δ = m	δ = mm
$\pi = 3.14$	$\pi = 3.14$	ρ = μΩ-cm
f = Hz	f = Hz	f = Hz
σ = % IACS	σ = mhos/m	
$\mu = 4\pi \times 10^{-9}$ *H*/cm	$\mu = 4\pi \times 10^{-7}$*H*/m	$\mu = 1.0$

Brown	P.S. 21207 MIL-STD-1537	Metals Handbook
$\delta = 1.98(\rho/f\mu)^{1/2}$	$\delta = 26/(f\mu\sigma)^{1/2}$	$\delta = 1980(\rho/f\mu)^{1/2}$
δ = in.	δ = in.	δ = in.
$\mu = 1.0$	$\mu = 1.0$	$\mu = 1.0$
f = Hz	f = Hz	f = Hz
ρ = μΩ-cm	σ = % IACS	ρ = Ω-cm

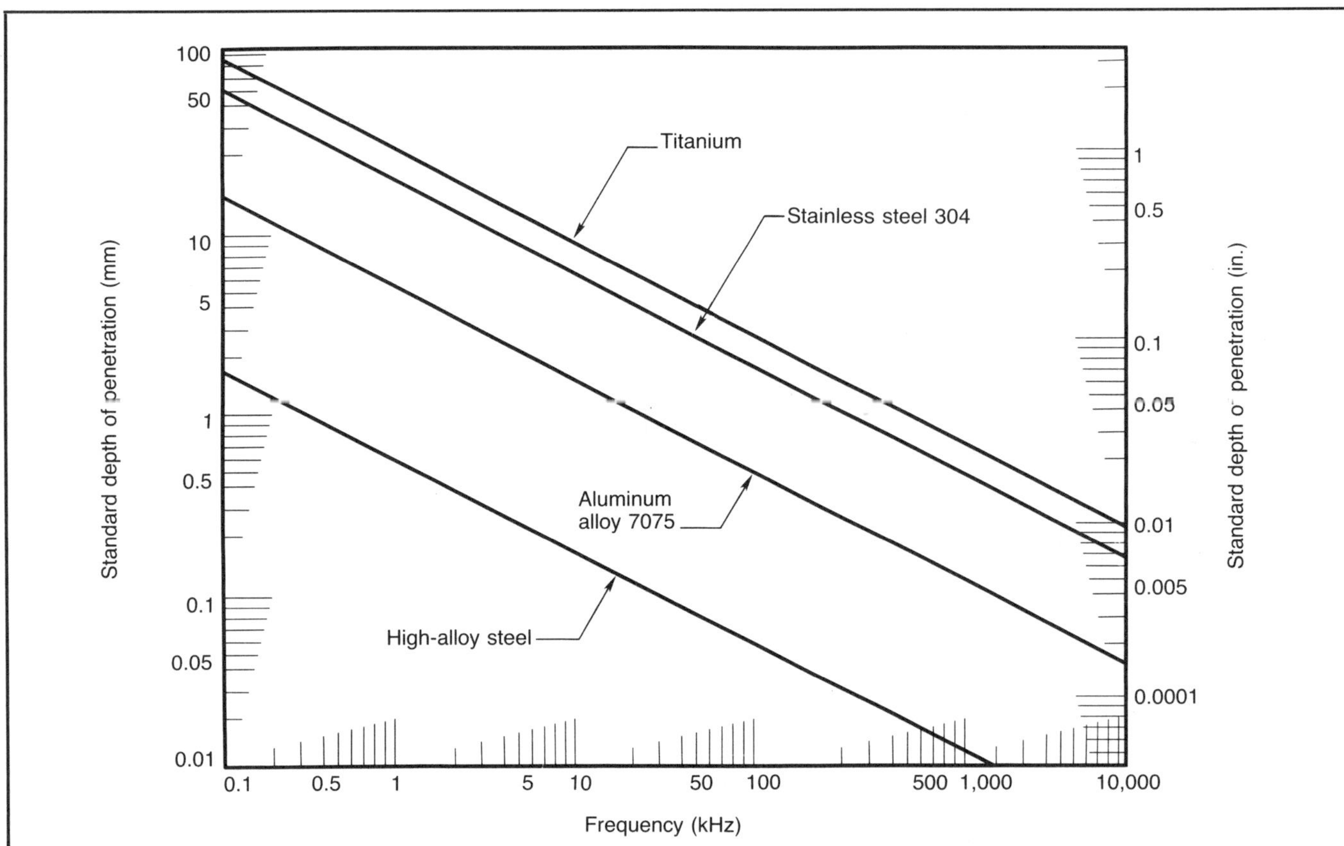

Figure 7-9. Eddy current standard depth of penetration for various-conductivity materials at different frequencies.

Chapter 8
Effects of Magnetic Coupling with Test Materials

Effect of Spacing between Test Coil and Test Materials

In Chapter 7, the interactions between eddy current test magnetizing coils and test materials were described in terms of simple two-winding transformer theory; the theory applies directly when the test-material surface is very close to the winding turns of the magnetizing coil. Because many practical eddy current test systems are used with some spacing between the coil windings and the test material (so that test objects can be readily handled and moved within the coil fields), the effect of such spacing upon induced eddy currents, and the reaction of the eddy currents upon the coil windings, are now discussed.

With eddy current probe coils, this spacing is described as the *probe-coil liftoff*. With encircling test-coil systems, the spacing between the magnetizing-coil winding and the surface of the test object is called the *coil fill factor*. In each case, the effect of increasing the spacing between the exciting coil and the surface of the test material acts to (1) reduce the magnitude of induced eddy currents and (2) reduce the sensitivity of the eddy current test to material property variations or discontinuities.

Effect of Probe-Coil Liftoff on Test-Coil Reactance

When an eddy current probe coil is lifted away from the surface of a nonmagnetic, conducting test material by some distance k, as shown in Fig. 5-3(a), a portion of the magnetic flux created by the test-coil current fails to reach the test material. If the probe coil is lifted so far above the test-material surface that none of its magnetic flux reaches the test material, the probe coil exhibits its empty-coil inductive reactance, X_{L_0}. This is the highest value attainable during tests of nonmagnetic materials. If the probe coil then approaches the surface of the test material, more and more of its magnetic flux lines intercept the test material, inducing eddy currents that oppose the coil's magnetizing field. As the eddy current reaction-field strength increases, the total magnetic flux linkages with the probe magnetizing coil are reduced.

As the eddy current reaction field increases with close proximity of the probe coil to the test-material surface, the probe coil's inductance and inductive reactance are accordingly reduced. The limit of this reduction is attained when the face of the probe-coil assembly is placed in firm contact with the test-material surface. The changes in probe-coil test signals (as the probe is lifted away from direct contact with the nonmagnetic test material) produce what is called the *probe liftoff effect*.

Effect of Probe-Coil Liftoff on Test Sensitivity

The reduction in eddy current magnitudes, as the probe magnetizing coil is lifted away from the surface of a nonmagnetic conducting test material, causes the probe-coil test signal to approach the empty-coil point (point P_0 on the complex-plane diagrams of Fig. 7-2). The magnitude of the test-signal variation diminishes, and the eddy current test sensitivity consequently decreases dramatically as the probe liftoff distance exceeds about one-tenth of the probe-coil diameter. Even with close spacing between the face of the probe coil and the test material, small spacings (in the range of 0.03 mm [0.001 in.]) are readily detectable. Thus, to attain reproducible test signals in probe-coil eddy current tests, it is essential to control the probe liftoff.

Because eddy current test sensitivity to test-material properties is greatest when the eddy current ohmic resistance losses are maximized, it is evident that maximum probe-coil sensitivity is attained with the probe coil in direct contact with the flat surface of a nonmagnetic test material. Liftoff always reduces the sensitivity of probe-coil tests to the properties of conducting test materials. Special phase-selective signal-analysis techniques have been developed to reduce or minimize the probe liftoff effect. See examples in Figs. 10-5 and 10-6.

Effect of Probe Tilting or Wobble

For reproducible eddy current tests, the axis (centerline) of the probe-coil winding should be perpendicular (normal) to the adjacent surface of the test material. During manual testing, precautions should be taken to place the face of the probe coil directly upon the flat surface of the test material to avoid the effects of probe tilt or wobble. Tilting of the probe coil moves at least a portion of the probe-coil winding away from the test-material surface and so increases the liftoff of a portion of the winding. Such liftoff reduces the coupling of the magnetizing coil with the test material and so reduces the eddy current test-signal magnitude. With mechanized probe scanning systems, it is equally essential to maintain the probe-coil axis perpendicular to the test-material surface and to maintain constant liftoff as closely as possible. Vibration of the probe mount, or of the test material, can result in nonrelevant test signals that may be confused with defect indications.

Effect of Rough or Curved Surfaces of Test Objects

Similar precautions are recommended during probe-coil tests of material with rough or curved surfaces. Surface roughness results in a varying degree of probe-coil liftoff, even when the probe-coil face is placed in direct contact with the raised asperities of the test-material surface. Such surface roughness is often encountered with unfinished castings or forgings. In this case, control or suppression of probe liftoff effects is desirable for maintaining reproducibility in test measurements. Phase-selective signal analysis techniques and modified test-coil designs have been used for this purpose.

Probe-coil tests of cylindrical test objects are feasible with orbiting or spinning probe-coil systems, where the probe-coil axis is kept normal to the surface of round bars or tubes. Where round bars or tubes are rotated while passing a fixed probe coil, guidance is necessary to avoid eccentric rotational movements as the test objects move past the probe coil. Unique systems have also been developed to permit probe-coil systems for scanning square or rectangular billets or shapes.

In manual probe-coil tests of test objects with curved surfaces, it is often helpful to provide a matching curved face (or shoe) on the probe assembly. The shoe fits the test-object curvature and ensures that the probe axis is perpendicular to the surface at the coil center. Alternatively, it is possible to design special probe coils with windings shaped into curves that fit closely to fillets or rounded corners of test objects. Shaping the probe-coil winding to fit test-object contours tightly ensures minimum liftoff, providing tight magnetic coupling of the entire coil winding with the test material.

Effect of Coil Fill Factor on Encircling Test-Coil Reactance

In encircling-coil eddy current tests, the coil winding is a solenoidal winding on a hollow coil form. Its inner annulus is large enough to permit round bars, tubes, or other cylindrical test objects to be placed within, or moved through, the magnetizing and signal-pickup coil fields (see Fig. 8-1). The maximum magnetizing field strength is, as always, directly adjacent to the magnetizing-coil winding turns. (The field along the central axis of the solenoidal coil is of somewhat smaller magnitude.) If a cylindrical test object is fitted very tightly within the magnetizing-coil winding turns, the magnetic field coupling with the test material would also be maximized.

However, if the test-object diameter is only a fraction of the magnetizing-coil diameter, the

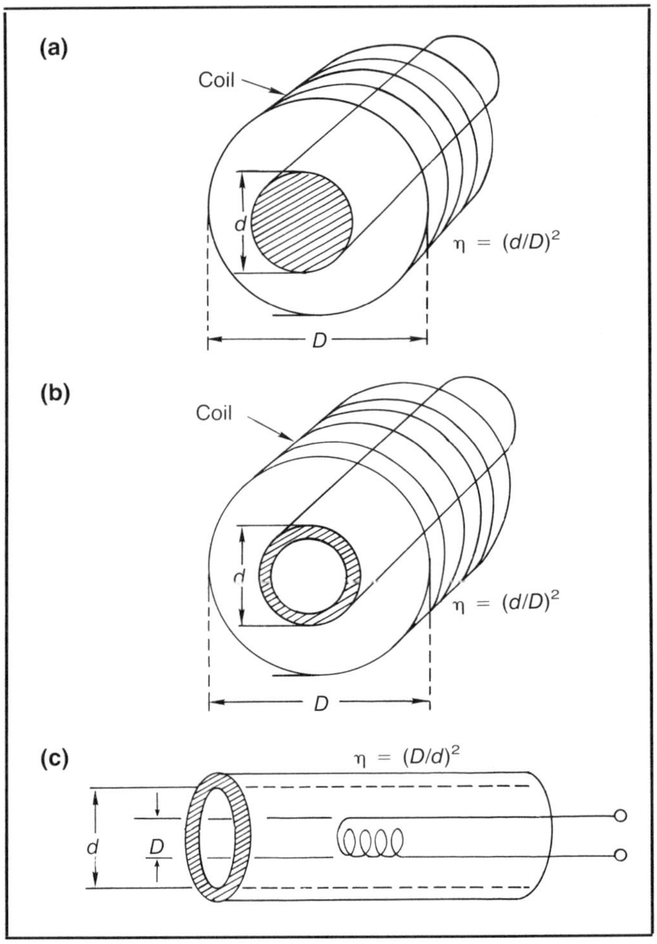

Figure 8-1. External and internal eddy current test-coil fill factors. (a) Test coil combining solid cylinder of smaller diameter *d*, partially filling the coil. (b) Test coil combining hollow cylinder of smaller diameter *d*, partially filling the coil. (c) Test-coil inside cylinder (fill factor still ≤ 1.0).

magnetic coupling is reduced because many magnetic flux lines could return through the air space between the outside surface of the test material and the winding turns. Because these flux lines, with paths wholly in air, do not intercept the test material, they are wasted and do not serve to induce eddy currents in the test objects. They contribute to the magnetizing-coil inductance and to pickup-coil voltage signals, but they do not measure properties of the test material. However, they do serve to measure the outside diameter of cylindrical test objects, and encircling-coil eddy current tests are often used for this important dimensional measurement. Their contribution to the eddy current test signal is totally a reactance effect because they do not induce any eddy current losses.

When the test object is significantly smaller in diameter than the encircling-coil winding turns, a much reduced portion of the test coil's magnetic flux interacts with the test material. The induced eddy currents are of smaller magnitude, and their reaction effects upon the test coil are diminished. This results in a lowering of the sensitivity for encircling-coil eddy current tests. The smaller the test-object diameter, the poorer the test sensitivity. A very fine, metallic wire at the centerline of a large-diameter test coil would produce negligible test-signal variations. Generally, it is desirable to fill the coil annulus as completely as possible to obtain maximum sensitivity in encircling-coil tests; encircling-coil eddy current tests can best be made with a large coil fill factor.

Definition of Coil Fill Factor

In practical *encircling-coil tests,* the coil winding inner diameter D must be considerably larger than the test-object outer diameter d so that test objects can pass freely through the coil assembly without contacting it at high speeds. The coil fill factor η (Greek letter eta, lowercase form) is defined by Eq. 8-1:

$$\eta = \left(\frac{d}{D}\right)^2 \qquad \text{(Eq. 8-1)}$$

Fig. 8-1 illustrates this relation. If the test bar or tube outer diameter were equal to the test-coil inner diameter, the coil fill factor η would equal unity (100 percent). If the test bar or tube diameter were only one-half of the coil winding diameter, so that the test object filled only one-fourth of the space within the coil, the coil fill factor would be 0.25, or 25 percent. The coil fill factor is proportional to the fraction of the coil area filled by the concentric cylinder of test material. The complex-plane locus curves are reduced in size and diameter as the coil fill factor becomes smaller (see Fig. 8-2). The eddy current test signal variations are also reduced in magnitude when tests are made with a coil fill factor smaller than unity.

There are numerous eddy current test applications in which the test coil is located within the test object. For example, large numbers of tubes are required for boilers, condensers, coolers, and heat exchangers in oil refineries, steam power plants, chemical plants, and nuclear plants. Such installations usually consist of large numbers of tubes located close together and enclosed in a vessel so that their external surfaces are not accessible. These tubes can be tested only by inserting the test coil into the tubes from available access points.

Another important problem is the internal testing of thick-walled tubes which, although their external surfaces are accessible, do not permit detection of internal surface defects because their heavy wall thickness limits the eddy current penetration depth. Fig. 8-1(c) illustrates a test coil of diameter D placed within a tube of inside diameter d. The fill factor η in this case can be calculated from Eq. 8-2:

$$\eta = \left(\frac{D}{d}\right)^2 \qquad \text{(Eq. 8-2)}$$

Precautions on Centering Test Objects within Encircling Test Coils

When testing round bars or tubes with encircling test-coil systems, it is important to provide sleeves or guides to assure that the test objects are centered within the encircling coil. If the test object is not centered, that portion of its surface closest to the coil winding will receive unduly high magnetization, whereas the portion of its surface

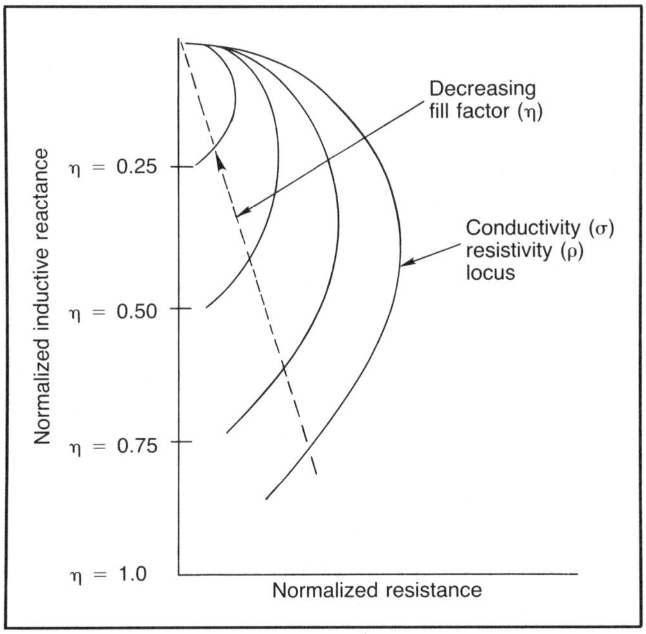

Figure 8-2. Normalized impedance-plane response for different fill factors, frequency, and resistivity of nonmagnetic bars and heavy-walled tubes.

farthest from the test-coil winding will receive less than normal magnetization. The test-signal response to off-center positioning of test objects is not linear, and the portion nearest to the coil winding exerts an unduly large influence upon the eddy current test signals. In some cases, signal effects from off-center positioning could be mistaken for variations in test-bar diameter. In other cases, they might indicate variations that do not actually exist in test-material properties. In any specific test application, it would be desirable to check the reproducibility of the test-object position and to control the positioning system so that there are no detectable effects on the signal from movements or vibrations of the test bars or tubes.

Precautions on Positioning Small Test Parts within Encircling Test Coils

When encircling-coil tests are made upon small test objects with various cross sections or special shapes (such as screw-machine products), it is essential to provide reproducible positioning of the test parts within the encircling coils. With short test objects, the ends of the parts must also be positioned precisely. If the short part is displaced longitudinally within the test coil, so that one end is within the coil field and the other end extends beyond the coil field, test signals will change as air replaces the conducting material. Because encircling-coil tests respond to the total eddy current field induced within the test material, irregular placement of successive test objects could lead to signal variations that might be mistaken for defects or for dimensional variations. For such tests, jigs or fixtures are commonly used to hold each test object in a precise location and orientation.

For small runs, the operator may place the successive test parts in these fixtures. In large-scale-production inspection systems, it is often feasible to feed parts into the fixture mechanically or to allow the test parts to fall through the encircling test coil. It is desirable in this case to provide a stop mechanism to hold each test part in the same position during the eddy current tests. Time gates are then used to make the eddy current test only when the part is in position (and usually stationary) and to eliminate transient signals that result when the small test parts are in motion. In such applications, photocells or sensing switches can be used to ensure that a test object is in the correct position when the measurements are made. (Some feeding systems for small parts occasionally fail to place a part within the test coil during each cycle of operation.)

Precautions on Selecting Materials for Test-Object Positioners

It is important to recognize that any conducting or magnetic materials placed within the magnetic field of an eddy current test coil can affect the eddy current test signals. For this reason, jigs, fixtures, and stops that are used to control test-object positioning in encircling-coil systems must be made of nonconducting, nonmagnetic materials (such as wood, organic resins, or laminates). In many cases, it is possible that even metallic screws used to assemble such fixtures could result in detectable eddy current signal variations. In particular, moving parts of such assemblies should be nonmetallic.

When fixtures are devised, it is a good precaution to determine whether or not any test-signal changes result when the fixtures are placed in the empty test coils. With elongated test objects such as rods or tubes, particularly of ferromagnetic materials such as steel, the coil field can extend for a considerable distance along the coil axis on either side of the coil winding. It should be ensured that no metallic materials within the field of the test coils can affect the test indications. Actual test measurements with and without the test-object fixtures in place can be used to detect any signal effects created by metallic elements other than the test objects.

Effect of Test-Coil Diameter on Reactance

As indicated by Eq. 7-7, the inductance of the test magnetizing coil is directly proportional to the total magnetic flux encircled by the coil windings. The *total magnetic flux* is the product of the flux density B and the area within the coil winding. The total magnetic flux increases in proportion with the square of the coil diameter D. For the empty test coil in air, the test-coil reactance (see Eq. 7-9) also increases in proportion to the square of the test-coil diameter. Small-diameter test coils have proportionately small reactance. Large-diameter test coils have much greater values of reactance. The empty-coil reactance also increases with the square of the number of series turns in the coil winding. Thus, the inductance and the reactance of the test coils are approximately proportional to the square of the product of diameter and number of turns:

$$X_{L_0} \propto D^2 N^2 \qquad \text{(Eq. 8-3)}$$

The voltages induced in a signal-pickup coil wound coincidentally with the test magnetizing coil would also vary in proportion to the square of the coil diameter and the square of the number of series turns in its winding. Fig. 8-3 shows the effect of changing the diameter of a probe coil while the operating frequency and material conductivity are kept constant. Increasing the coil diameter moves the impedance point down the curve.

Effect of Test-Coil Diameter on Projection of Magnetic Field in Air

The size of the air space surrounding a magnetizing coil, through which its magnetic field extends, is directly proportional to the diameter of the coil. With circular coils, for example, the pattern

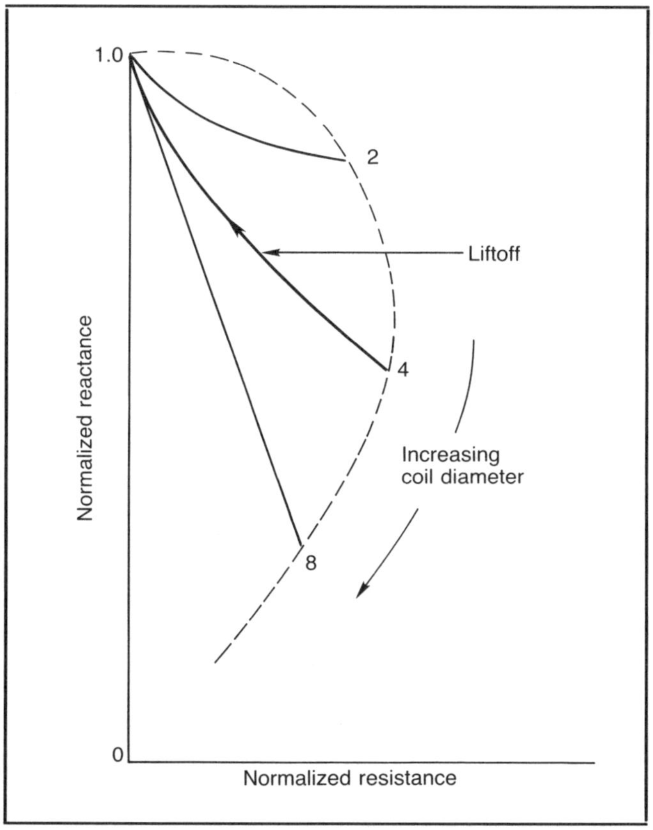

Figure 8-3. Impedance graph showing changes in probe-coil diameter.

of the magnetic-field flux distribution is similar in shape for all coils, but its dimensions are proportional to the circular coil's diameter. Small coils have magnetic fields confined to a small region around the coil winding. Large coils have magnetic fields that project into space farther away from the coil. The size of the magnetic field in space is directly proportional to the coil diameter *D*. Thus small-diameter coils measure test-material properties only in the region close to the coil. With large-diameter coils, a much larger area of the test-object surface is subjected to magnetization and to eddy currents induced by AC magnetization.

The diameter of the test coil also determines the size of its field of influence, within which external metallic materials could exert a change upon the test signals. Typically, this region of influence would extend at least three coil diameters away from the coil winding and, with highly sensitive eddy current test systems, to even greater distances. The extent of this region of magnetic influence may be determined experimentally by moving a ferromagnetic object through the air space surrounding the test coil and observing where its influence upon the eddy current test signals occurs. Conducting or magnetic jigs or fixtures within the field of influence should be eliminated so that they do not affect the eddy current test signals.

In mechanized systems where test parts are fed continuously to the test-coil station, the incoming and outgoing parts on either side of a test part must be sufficiently far from the test coil so that they do not influence the test indications. In addition, as either the front or back end of a long bar or tube enters the field of influence of the test coil, it can be expected that the test signals could change. In many test systems, special provisions to suppress end effects are employed. Otherwise, end-effect signals simulating those of discontinuities or property variations could actuate alarms or rejecting mechanisms even when good test materials are passing through the coil's field of influence.

Effect of Probe-Coil Diameter on Axial Projection of Magnetic Field

The relationship between the magnetizing-coil diameter and the size of its magnetic field in air applies also to the case of eddy current tests made with probe coils. Because the probe coil is typically laid flat upon the surface of the test material, the important factor is the distance to which its magnetic field projects along the probe-coil axis (or centerline of its windings). When the probe coil is in contact with the surface of the test material, this axial distance corresponds to depth within the test material. When the probe coil is spaced above the material surface, this distance includes the probe liftoff and the additional distance to which its magnetic field penetrates into the test material. The axial-field projection distance is directly proportional to the diameter of the probe magnetizing-coil winding. Small-diameter probe coils project their fields axially for very short distances. Large-diameter probe coils can project their magnetizing fields to correspondingly greater distances.

In general, the strong magnetizing field exists within an axial distance on the order of one-tenth of the coil diameter only. At an axial distance equal to one-third of the probe-coil diameter, the field is typically about 50 percent of its strength at the coil face. At an axial distance equal to the coil diameter, the magnetizing-field intensity is only 10 percent of that near the coil face. Such low magnetizing fields at such a distance from the coil winding would produce very small eddy current test signals. Most of the eddy current test signal is derived from (1) the strong magnetic field and (2) the intense eddy currents in the test-material layers closest to the face of the probe coil.

Use of Large-Diameter Probe Coils to Compensate for Large Liftoff

If the eddy current test conditions require large probe-liftoff spacing (as with orbiting probe-coil systems, which must provide adequate clearance for moving test materials), it may be necessary to use a larger-diameter probe-coil winding than would be used during contact tests. The large coil can project a reasonably strong magnetizing field to the test material and can ensure adequate levels of test signals. However, the area of the test material inspected by the larger probe coil at any instant is increased in proportion to the probe-coil diameter. This may reduce sensitivity to small discontinuities such as cracks whose lengths are only a fraction of the probe-coil diameter. However, to detect discontinuities or property variations in the test material, it is essential to provide magnetizing-field strengths sufficient to induce eddy currents with detectable reaction effects.

Use of Large-Diameter Probe Coils to Test Thick Nonmagnetic Plate

With ferromagnetic materials, or with high-conductivity nonmagnetic materials, the penetration depth of eddy currents is limited by the skin effect and eddy current attenuation. However, with thick layers of poorly conducting nonmagnetic materials, such as austenitic stainless steels, titanium or its alloys, or other nonmagnetic metals and alloys with low electrical conductivities, it is often possible to make eddy current measurements to considerable depth (10 to 30 mm or more). Where the standard depth of penetration is equal to perhaps half of the total material thickness, it is essential to use low test frequencies to attain sensitivity to total plate thickness and to subsurface and far-surface discontinuities.

However, in these applications, small probe coils may fail to project their magnetizing fields into the lower layer of the test material, so no eddy currents can be induced at these great depths. Larger-diameter probe coils (whose diameters equal or exceed the total test-material thickness) can project their magnetizing fields to the depths required for detection of overall thickness variations, root defects in fusion welds, or detection of internal defects. For example, for electron-beam welds in 6-4 titanium alloy of 25 mm (1 in.) thickness, good response was obtained from lack of fusion in the root areas by this technique. Semiconductor detectors of the eddy current signals were employed.

More generally, with thick specimens of graphite and other materials with conductivities in the range below 10 percent IACS, it is probable that the projection limitations of the test coil's magnetizing field, rather than the usual limitations of eddy current attenuation and skin effect, will control the depth sensitivity of probe-coil tests. In these special applications, it is appropriate to adjust the probe-coil diameter and to select suitably low test frequencies to improve depth resolution. By this means, eddy current tests can be extended to test-material thicknesses often considered impractical with conventional small-diameter probe-coil systems used with high-conductivity test materials.

Use of Small-Diameter Probe Coils to Test Thin Coatings

When the purpose of probe-coil eddy current tests is to measure the thickness or properties of thin coatings on a conductive base, small-diameter probe coils are usually preferred. These, in combination with suitably high test frequencies, can provide very high sensitivity to thin coating layers and minimize effects related to property or thickness variations in the underlying base metal. The purpose here is to confine the magnetizing field, and the resulting eddy current distribution, to the thin coating layer and to minimize the field within the base metals. It is also highly desirable to provide sensitivity to small discontinuities and defects in the coating layer, if these are significant to the test.

If the thin coating is nonconducting material, such as paint, porcelain enamel, or plastic, the basic measurement is that of liftoff. Sensitivity to probe-coil liftoff through small distances is increased as probe-coil diameter is made smaller. Test coils of 3 mm (0.13 in.) or smaller diameters can be employed. Small probe coils with ferrite cores to concentrate the magnetic field at their tips find wide acceptance.

If the thin coating is a conducting material (such as silver, gold, cadmium, aluminum, tin, lead, or zinc plating) whose conductivity is higher than that of the base material, small-diameter probe coils and high test frequencies can be employed to concentrate most of the eddy currents within the coating layers. This can provide high sensitivity to the thickness of the conductive coating and, in some cases, to the degree of alloying or impurity content of the coating alloys. Some commercially available eddy current coating-test instruments provide calibrated probes and scales of coating thickness or conductivity and guides for test-frequency selection for each specific application.

Chapter 9
Coils, Instruments, and Standards

Introduction

Eddy current inspection appears complicated to the novice because of the numerous types, sizes, and shapes of test coils, instruments, and instrument-calibration reference standards. In this chapter, the author will try to simplify and clarify the types and selection of the previous items.

Types of Test-Coil Systems

Fig. 9-1 shows three basic coil systems used for inspecting small-diameter tubing. These are

1. encircling, or "feed-through," coils,
2. inside probe, or "bobbin," coils, and
3. surface probe coils.

Also illustrated for each case is the use of both "absolute" measurement (without a direct reference) and "differential" methods. The *encircling-coil* system is more common because of its ease of use and relatively high inspection speeds. The mechanical difficulties involved in the application of *bobbin coils* are considerable, particularly in the case of long lengths of very small-diameter tubing. Despite this, their use may be warranted for inspection of the bore of heavy-wall tubing, duplex tubing, or heat-exchanger tubing. The surface coil is not in general use for inspection of small-diameter tubing because of inherent mechanical problems and slow inspection speeds. Its high sensitivity to small discontinuities and definitive ability, however, are very attractive for critical inspection.

The use of *single-* and *double-coil arrangements* is also illustrated in Fig. 9-1. In the case of the double encircling-coil system, the AC excitation is applied to the outer coil, and the voltage induced in the inner "pickup" coil is examined. The advantage of this arrangement is that the pickup coil may have a very small dimension in the longitudinal direction and thus be sensitive to very short discontinuities. At the same time, the exciting field in the vicinity of the pickup coil may be made almost a geometrically perfect cylinder of flux parallel to the axis of the coil and with uniform density in the absence of the tube. This is done by the use of an elongated exciting coil. The double-coil system does not appear to offer a great practical advantage over the single-coil system with a large ratio of outside diameter of tube to effective coil diameter because the effect of field distortion in the vicinity of the tube wall produced by the use of a thin single coil is small.

From the standpoint of interpretability, the probe-coil system (Fig. 9-1) is very attractive. In such a system, the probe mechanically scans the tube surface in a helical pattern. The advantage of the probe coil is its inherent ability to be definitive. Because it is a surface probe, it is not affected by changes in diameter. Changes in wall thickness are indicated by changes in the signal level, which is a function of the position along the tube but is independent of the tube rotation. Eccentricity is indicated in the signal level, which varies slowly as a function of rotation. Discontinuities within the wall of the tube, because of their abrupt nature, cause sharp indications in the pattern.

The difficulty with the probe-coil system is that even with a very carefully designed coil-holding mechanism, wobble of the tubing does occur, and the signal varies with the coil-to-tube spacing. These signals sometimes look like those obtained from cracks.

Classification of Feed-Through Coil Arrangements

Various coil arrangements are used in eddy current testing where cylindrical test objects are inserted into the coil for measurement. Coil types include

1. single primary coil, Fig. 9-2(a),
2. primary and secondary coils, Fig. 9-2(b),
3. bridge arrangement of two primary coils, Fig. 9-2(c),

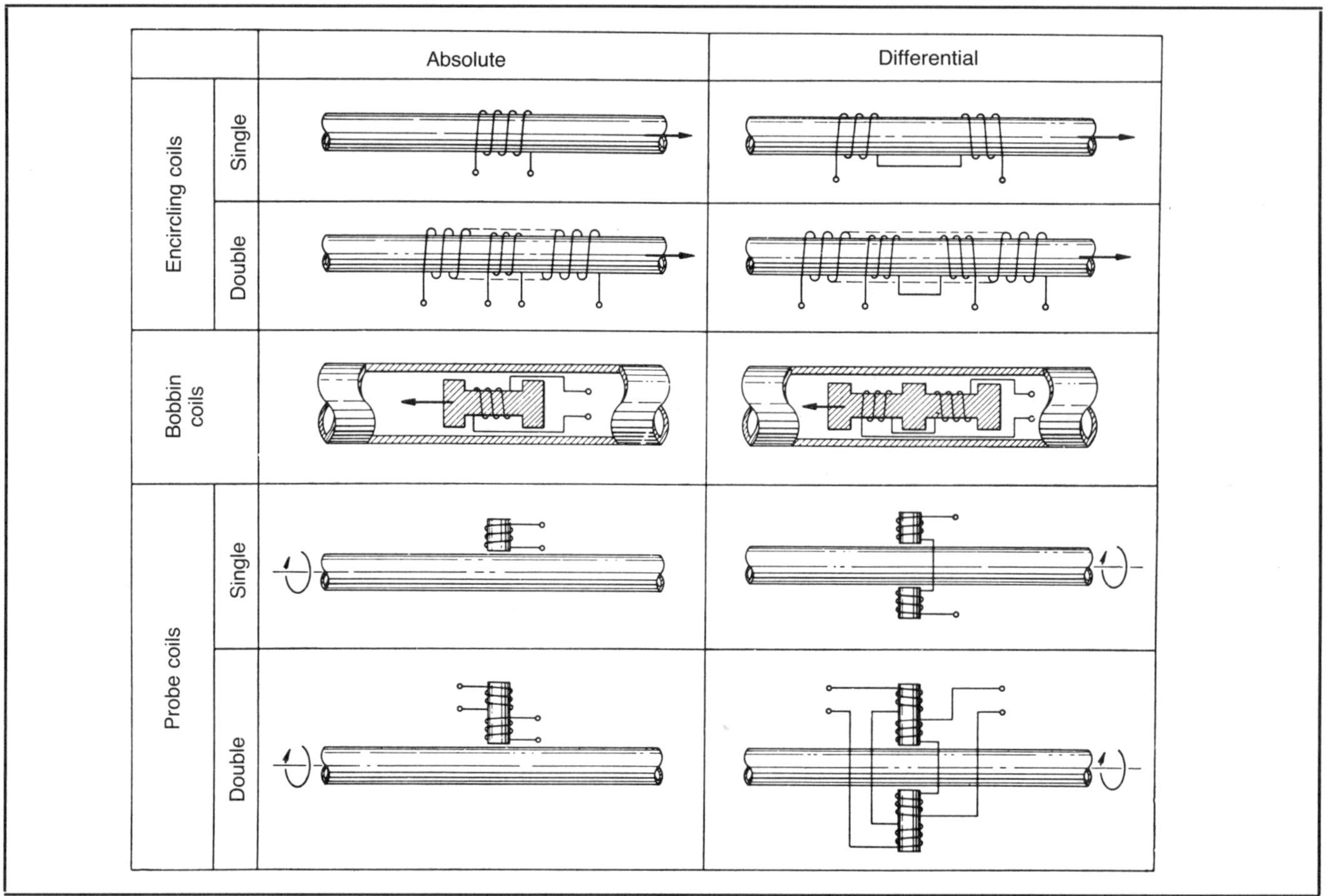

Figure 9-1. Test-coil configurations for eddy current testing of small-diameter tubing.

4. bridge arrangement of two primary and two secondary coils, Fig. 9-2(d),
5. comparison arrangement of two primary coils, Fig. 9-2(e), and
6. differential arrangement of two primary and two secondary coils, Fig. 9-2(f).

With these various coil arrangements, quantitative measurements can determine

1. the absolute value of the electrical conductivity σ and
2. phase and amplitude variations caused by diameter or conductivity variations or by a surface crack of specific depth.

These quantitative measurements are possible even if the coil data (fill factor η) or the electrical conductivity σ of the test cylinder are unknown.

Fig. 9-2(a) shows the very simple case in which the test coil consists of a single primary winding. In this case, the apparent impedance of the empty coil or the complex voltage across the empty coil changes when the test cylinder is inserted, as previously shown in Fig. 5-3(a) and (b). A similar reaction is obtained for the primary and secondary coil arrangement of Fig. 9-2(b).

In the bridge arrangement with two primary coils, Fig. 9-2(c), or two primary and two secondary coils, Fig. 9-2(d), no voltage appears across the bridge circuit or across the differentially connected secondary coils in the absence of the test cylinder. The bridge is balanced, or the two secondary coil voltages cancel each other in the "empty state." When the test cylinder is inserted into either of these coil arrangements, the voltage changes. If another test cylinder with the same physical properties is now inserted into the second (empty) coil, the voltage across the bridge terminals, or at the output of the differentially connected secondary coils, disappears. If the physical properties (for example, the conductivity) of one test specimen differ from those of the comparison specimen, their effective permeabilities will differ also. In this case, a voltage difference will appear across the terminals of the bridge of Fig. 9-2(c) or the secondary coils of Fig. 9-2(d). This voltage difference ΔE will be related to the difference in effective permeabilities.

In the bridge arrangements of Fig. 9-2(c) and (d), the second test cylinder served only for compensation of the fundamental voltage. The test-coil arrangements of Fig. 9-2(e) and (f), on the other hand, employ a different portion of the same cylindrical test object as a *standard of comparison.* This is often called a "self-comparison" arrangement. In this case, a voltage appears at the output terminals of the coils only if the local effective permeability of the test cylinder differs at the locations of the first and second test coils. If the cylinder is inserted into only one of the two coils, the voltage will again appear at the coil output terminals.

The "self-comparison" arrangement is often used in crack-test instruments because the crack depth normally varies from location to location along the test cylinder. (Note, however, that a long, uniform crack is indicated only at its ends and not

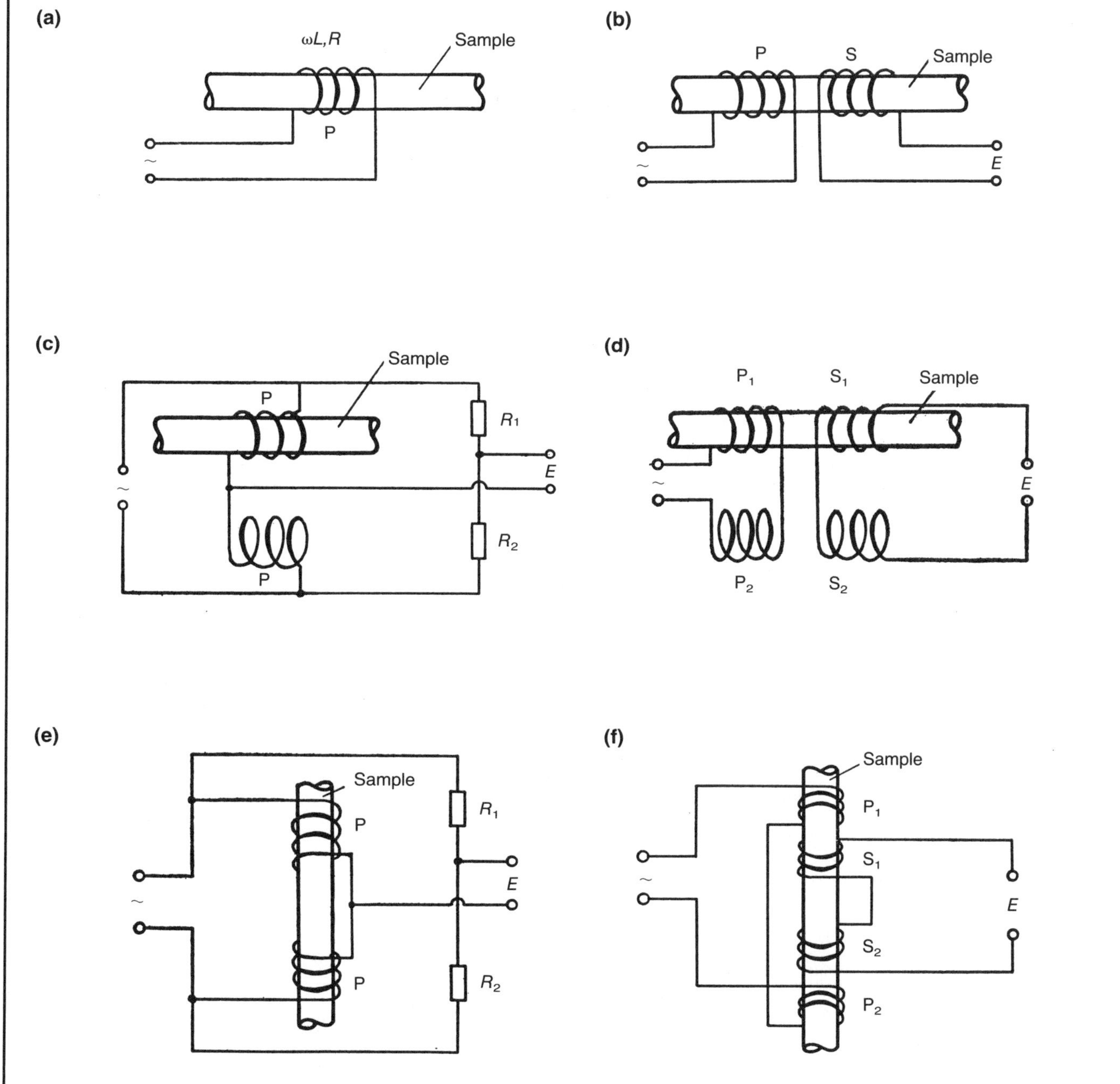

Figure 9-2. Arrangements for test coils for feed-through tests. (a) Single primary (P) test coil. (b) Primary and secondary (S) coils. (c) Bridge circuit of two primary coils. (d) Bridge circuit with two primary and two secondary coils. (e) Self-comparison arrangement of two primary coils. (f) Self-comparison arrangement of two primary and two secondary coils.

along the uniform, continuous portions.) Variations in physical properties, such as conductivity, relative magnetic permeability, and diameter, from one test cylinder to another have no effect with these coil arrangements because their influence on the effective permeability is compensated by the differential connections of the double test coils. Slowly varying physical properties in the test cylinders are also suppressed if the two self-comparison coils are close to each other. The fact that cracks of constant depth cannot be indicated is a limitation of the self-comparison test method.

Eddy currents produced by test coils always flow parallel to the plane, or normal to the axis, of the coil producing them. The eddy current field is strongest near the surface of the material, and its strength decreases with depth. The width of the coils may be varied as a function of test application (Fig. 9-3). Wide coils cover large areas, so they respond mostly to bulk effects: conductivity, relative permeability, diameter, etc. Narrow coils cover a smaller area and therefore are more responsive to small changes such as are produced by cracks or pits.

Surface Probe Coils

Coils for surface inspection may be inserted in a large variety of housings. The coils may be absolute or differential, and some may have a cylindrical shield outside the coil ("shielded coil") made from mu-metal, ferrite, or steel. These shields restrict the magnetic field and help to eliminate edge effects. Typical eddy current surface-probe configurations are illustrated in Fig. 9-4. These probes may have ferrite cores, used to cause a greater magnetic flux and eddy current field.

The pencil-point probe is generally used to inspect concave radii of test parts. The right-angle pencil-point probe is generally more convenient to handle than the straight probe. The bolt-hole probe is inserted into holes in structure, after removal of the fastener, for detecting cracks or corrosion pits. The above-mentioned probes generally have small coils and operate at high frequencies. Flat-surface probes are used as the name implies—on parts having flat surfaces. This type of probe may be spring-loaded to allow constant pressure between the probe and part so as to reduce liftoff effects. Flat-surface probes are generally found on conductivity and plating-thickness instruments.

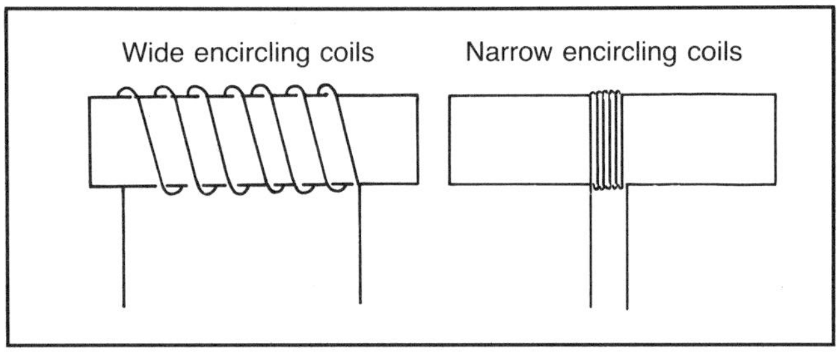

Figure 9-3. Wide and narrow encircling coils.

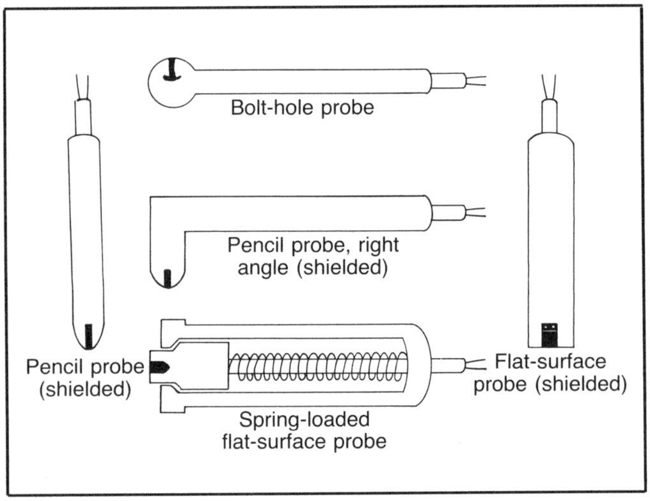

Figure 9-4. Typical eddy current surface-probe configurations.

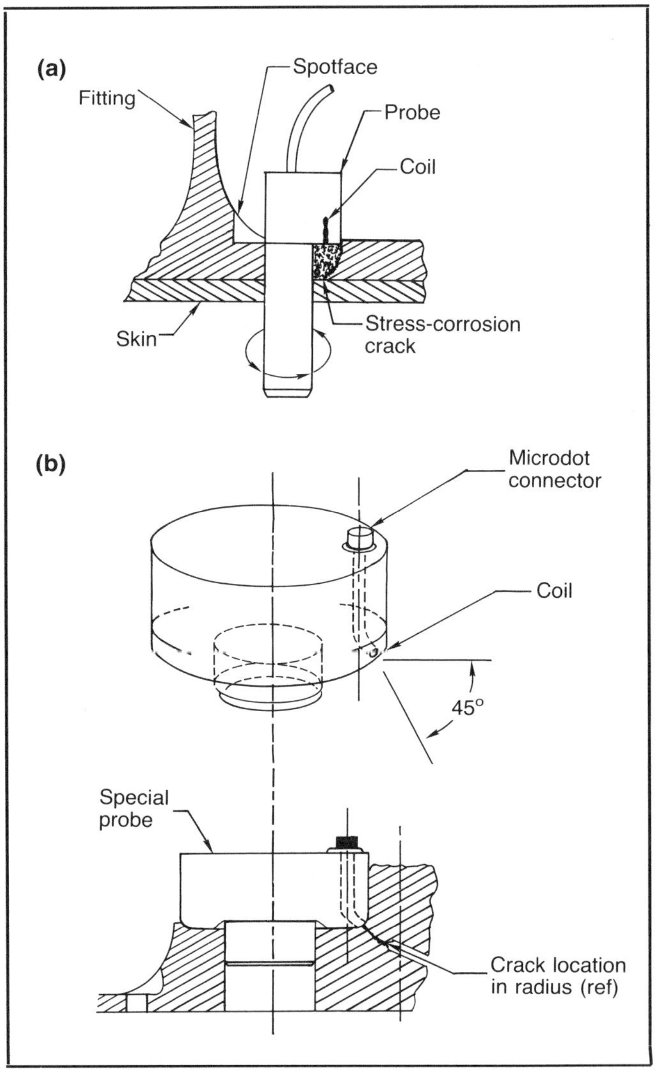

Figure 9-5. Two types of eddy current plug probes. (a) Plug probe used to detect crack at edge of fastener hole. (b) Plug probe used for detecting crack in part radius.

Structures generally develop fatigue cracks or stress-corrosion cracks at fastener holes or in change-of-thickness radii. For in-hole cracks, the bolt-hole probe is used. However, when it is desired to detect cracks at an exposed surface adjacent to the hole or in a spotface radius, the *plug probe* is generally used (see Fig. 9-5). For detection of subsurface cracks in aircraft structures, *encircling "donut"* and low-frequency, large-diameter, flat-surface probes have been used. Fig. 9-6 illustrates this application and shows the encircling donut probe. The central portion of the donut probe is hollow so that the center of the coil may be aligned with the axis of the fastener. The flat-surface probe must be placed between fasteners; the donut probe must be placed over the center of each fastener.

Because of the problems associated with using encircling donut and flat-surface probes for detecting cracks adjacent to aircraft fastener holes, the sliding *"driver/receiver-"* or *reflection*-type probes were developed (see Fig. 9-7). For inspection of aircraft with installed, flush-head fasteners, better crack sensitivity and reliability are obtained when using the sliding (driver/receiver)-probe system. In addition, these probes exhibit a wide frequency range and provide good penetration.

The major problem associated with probes is that each instrument manufacturer uses different electrical connectors. Hence, probes from different sources must have compatible connections to be used with each instrument. Often, adapters or "pigtails" are used to allow connection when probes from one manufacturer are used with instruments from another manufacturer. Also, some manufacturers use a second balancing coil in the probe housing; others have the balancing coil within the instrument. To avoid problems, when ordering probes (from a manufacturer other than the one who made the instrument being used), specify the instrument to which the probe will be connected.

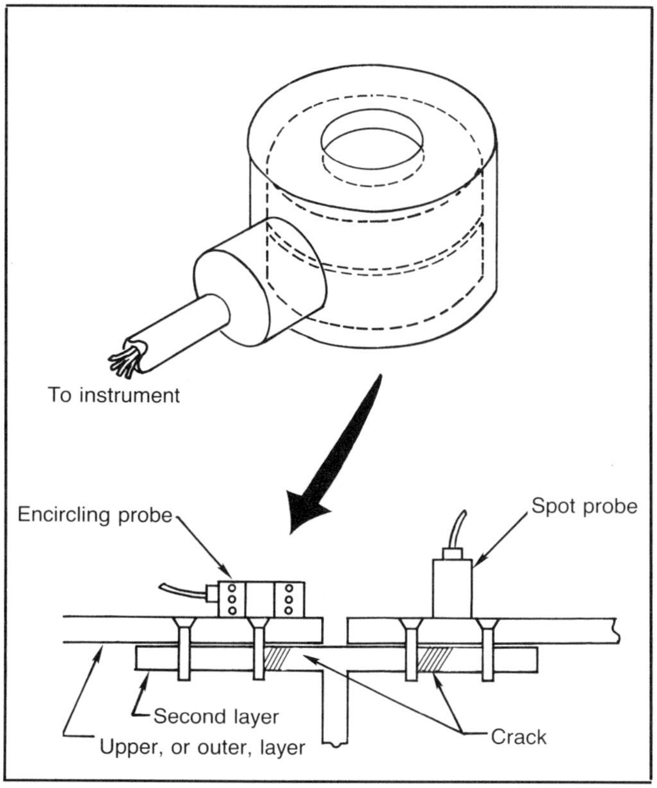

Figure 9-6. Detection of cracks using low-frequency eddy current encircling probe.

Eddy Current Instruments

Probably one of the most confusing aspects of eddy current testing is the variety of instruments available on the commercial market and the choices from which one must select the proper one for an inspection problem. Generally, eddy current instrumentation can be broken down to four basic classes:

1. Crack detectors.
2. Plating thickness instruments.
3. Conductivity meters for measuring conductivity in nonferrous alloys.
4. Multipurpose instruments that can be used to perform functions indicated in 1, 2, and 3 above.

The theory developed and presented in the preceding sections on eddy current tests provides data on the optimum test conditions, test frequency, variations in measured quantities, suppression of undesired influences, and related parameters for various test problems in the field of eddy current testing. However, many eddy current test instruments were developed on an empirical basis before present-day theory was available. Consequently, their designs deviate considerably from the optimum indicated by theory for specific test problems. Fortunately, the theoretical work of Dr. Förster led to the development of instruments, at his laboratory in Germany, that provided the necessary tools for performing eddy current tests to solve specific problems.

Electromagnetic testing in the last 110 years has evolved from relatively simple devices for metal

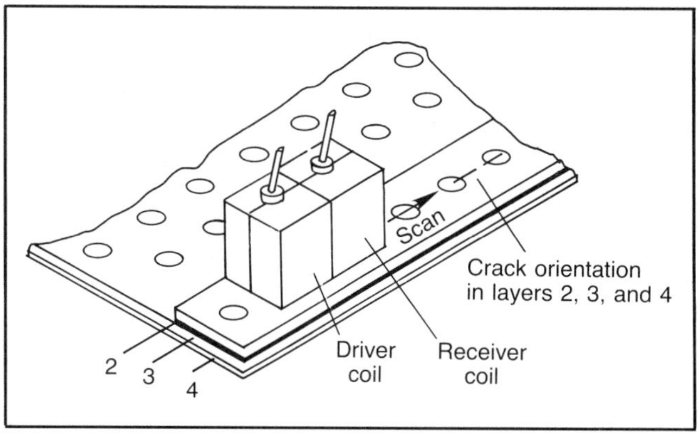

Figure 9-7. Sliding "driver/receiver" reflection probe used to detect subsurface cracks at fasteners in aircraft structure.

characterization to microwave testing and sophisticated systems with quadrature phase analysis. Active practical interest in electromagnetic methods for sorting metals and detecting discontinuities did not result in many useful test devices before the beginning of the 20th century. Numerous electromagnetic induction or eddy current comparators were patented in the United States in the period from 1925 until the end of World War II in 1945. For the most part, many of these early comparator systems were short-lived and received little acceptance in industry. The introduction by Förster of sophisticated, stable quantitative test equipment and practical methods for analysis of quantitative test signals on the complex plane were by far the most important factors contributing to the rapid development and acceptance of electromagnetic induction and eddy current tests during 1950–65 in the United States.

In the United States, numerous facilities began research to test these new concepts and instrumentation. Significant efforts were made at Oak Ridge, Hanford, and other facilities.

A basic eddy current instrument consists of an oscillator, a modified Wheatstone bridge (of which the sensor coil is a part), an amplifier, and a read-out device (such as a meter), as shown in Fig. 9-8.

The signal generated by the oscillator is applied to the sensor coil. This sets up an alternating magnetic field in the coil. The magnetic field will generate eddy currents in any conductive material in proximity. The eddy currents will generate a secondary magnetic field, which induces a current in the sensor coil. This causes a change in the impedance of the coil. The magnitude, the phase, and the rate of this change depend on both the instrument parameters and the conditions of the test part. If instrument parameters are controlled, then variations in the test part may be quantified.

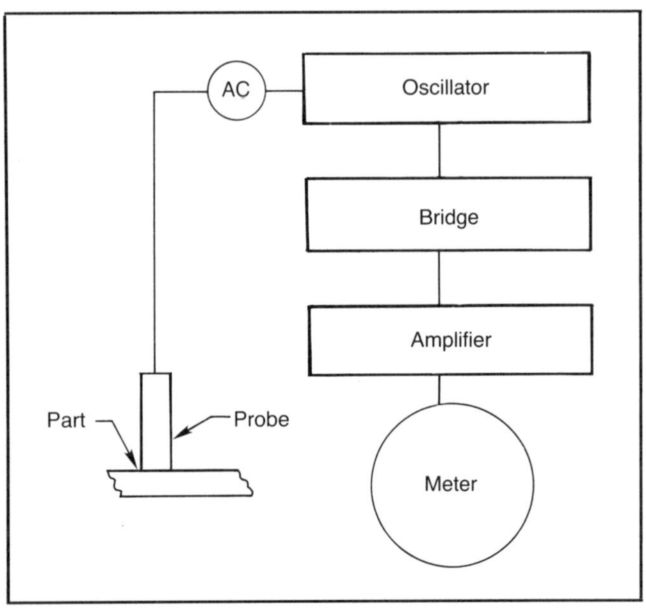

Figure 9-8. Simplified eddy current instrument.

The many different design variations are indicated in the classifications of eddy current instrument characteristics, which are discussed below.

Types of Test Coils

The most common types of test coils used in eddy current equipment were discussed in previous chapters, but are reiterated for continuity:

1. *Feed-through coils,* in which test objects such as rods, tubes, extruded or rolled shapes, wires, and other parts are passed through a hole within the test coil.
2. *Inside test coils,* which are inserted within cavities such as the interiors of hollow tubes or drilled holes.
3. *Probe coils,* which are placed on the surfaces of test objects.
4. *Forked coils,* through whose arms test objects such as sheets, foils, metallized paper, and similar conducting materials pass during tests.

Test-Coil Circuit Arrangements

Test coils are connected to the electronic test instruments in various circuit arrangements, which produce varying measurement effects, as follows:

1. Resonance-circuit methods, in which the test coil is part of a resonant circuit within the test instrument.
2. Time-differentiated resonance methods, in which a resonance coil is connected to a circuit with time differentiation to suppress slowly changing variations in the test objects.
3. Differential arrangements with a comparison coil (called the "absolute coil"). In these, the measurement voltage from the coil containing the unknown test object is compensated by that from a second coil containing a standard test object.
4. Time-differentiated comparison arrangements, using comparison coils, but with time differentiation in the indicating circuit to suppress *slow* variations in the test object.
5. Self-comparison differential coil methods, in which one portion of the test object is compared with another portion of the same object.
6. Time-differentiated self-comparison methods, in which the output of a differential coil system is subjected to time differentiation in the circuit, so that only effects with a rapid time variation of the impedance difference are indicated.

Type of Presentation of Test Indications

Various methods are used to present eddy current test indications to the observer or to automatic selection or control devices. Among these are the following:

1. Direct meter indication with pointer-type instruments.
2. Meter indication with phase-controlled rectifier.
3. Cathode-ray-tube (CRT) indicators, showing a generalized representation of the test indication (see Fig. 9-9).
4. CRT point or "flying-dot" method, in which the apparent impedance plane is represented by the screen of the CRT and measured values appear as luminous points or phasors on the screen (see Chapter 10). This method is the most popular at the present time.
5. CRT ellipse method, in which the measurement effect reacts on the vertical deflection plates of a CRT, and a reference voltage, whose direction on the impedance plane can be selected at will, is applied to the horizontal deflection plates (see Fig. 9-10).
6. Linear time-base method, in which the measurement effect influences the vertical deflection of the CRT beam. Here a linear sawtooth timing voltage, whose phase position can be selected at will, is applied to the horizontal deflection plates (see Fig. 9-11).
7. Time-slit selection method, in which a specific instant or phase-angle point is selected in the sinusoidal signal or control voltage, and measured values at that time instant only are indicated or evaluated (see Fig. 9-12).

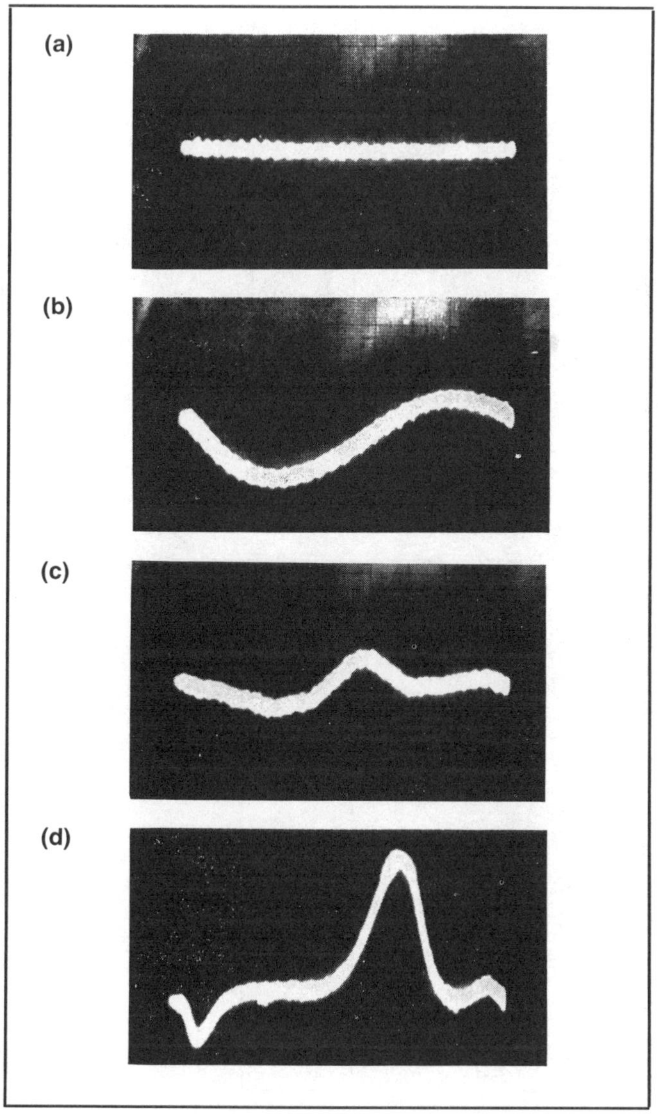

Figure 9-9. Probe-coil eddy current B-scan photographs. (a) No indication of defects. (b) Indication of tube-wall eccentricity. (c) Indication of small defect. (d) Indication of large defect.

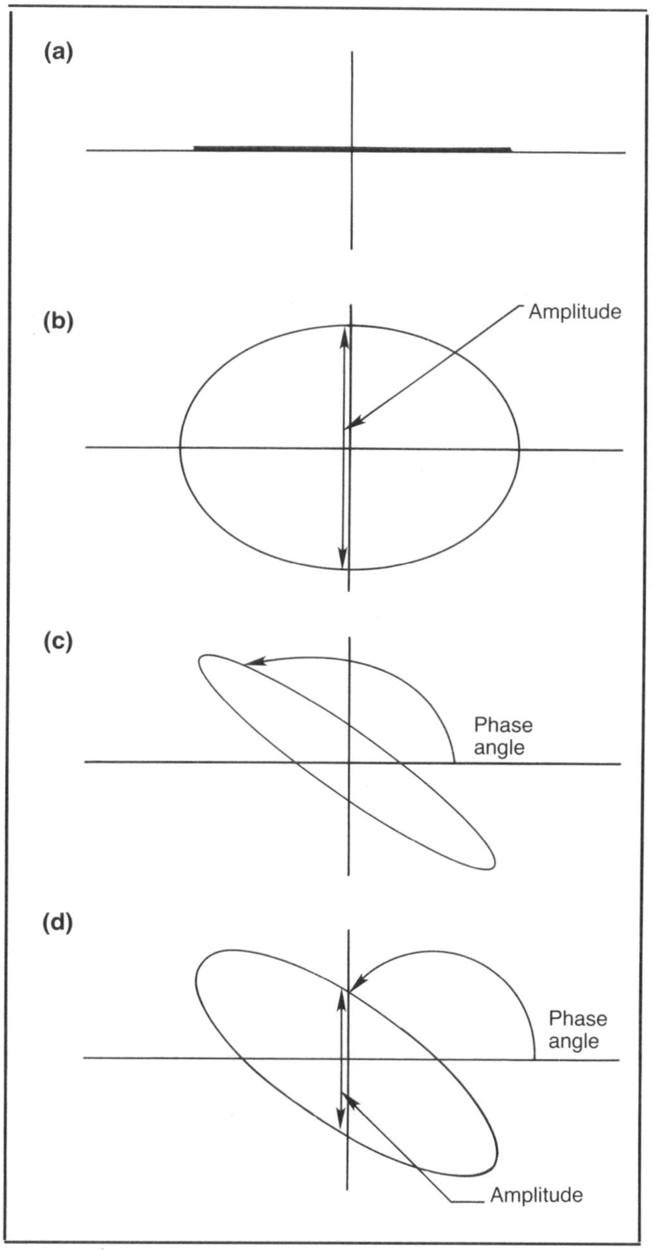

Figure 9-10. Ellipse (Lissajous) readout using phase discrimination. (a) Test part = standard (balance). (b) Crack in test part; diameter = standard. (c) Diameter change; no cracks. (d) Diameter change; with crack.

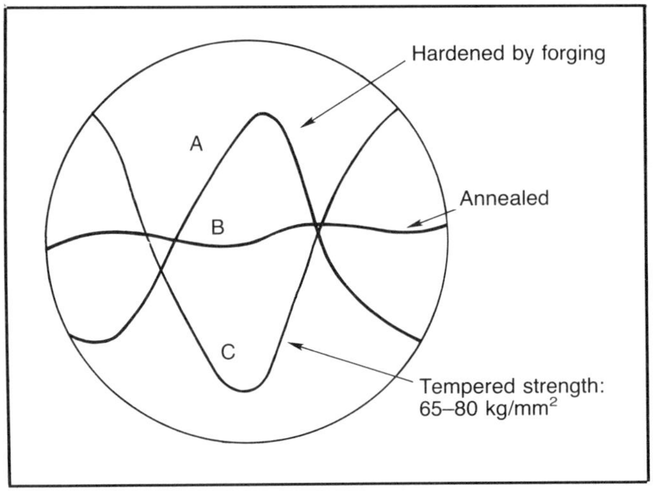

Figure 9-11. Linear time-base method showing different conditions for forged parts.

Electronic Factors Affecting Flaw-Detection Efficiency

Electronic analysis circuits and systems for eddy current flaw-detection systems should provide the following basic capabilities:

1. Sensitivity sufficient to produce usable signal indications under all test conditions.
2. Sufficient stability that readjustments of controls are not required frequently.
3. Sufficient selectivity that desired indications are always conveniently observable.
4. Reasonable simplicity of controls for initial setup, adjustments, and operation of instruments.
5. Speeds of response adequate to indicate all significant test signals, for all anticipated feed rates, of test-object materials.
6. Adaptability to test automation under existing as well as anticipated future conditions used for testing.
7. Compatibility with expected environmental conditions (such as severe noise, vibrations, electrical interference, dirt, scale, lubricants, extreme temperatures, and similar operating conditions).

Flaw-detecting system capabilities are determined by the following basic factors, which should be considered in system design:

1. Type of eddy current signal sensing system.
2. Geometry of the sensing system.
3. Test frequency to be used.
4. Signal phasing and phase limitation.
5. Signal filtering.
6. Saturating field requirements (for tests of ferromagnetic materials).

Depending upon the nature of the tests, test-object handling-equipment needs may be minimal or may require very complicated mechanical design. In some tests, the test-coil assemblies are designed so that they are positioned

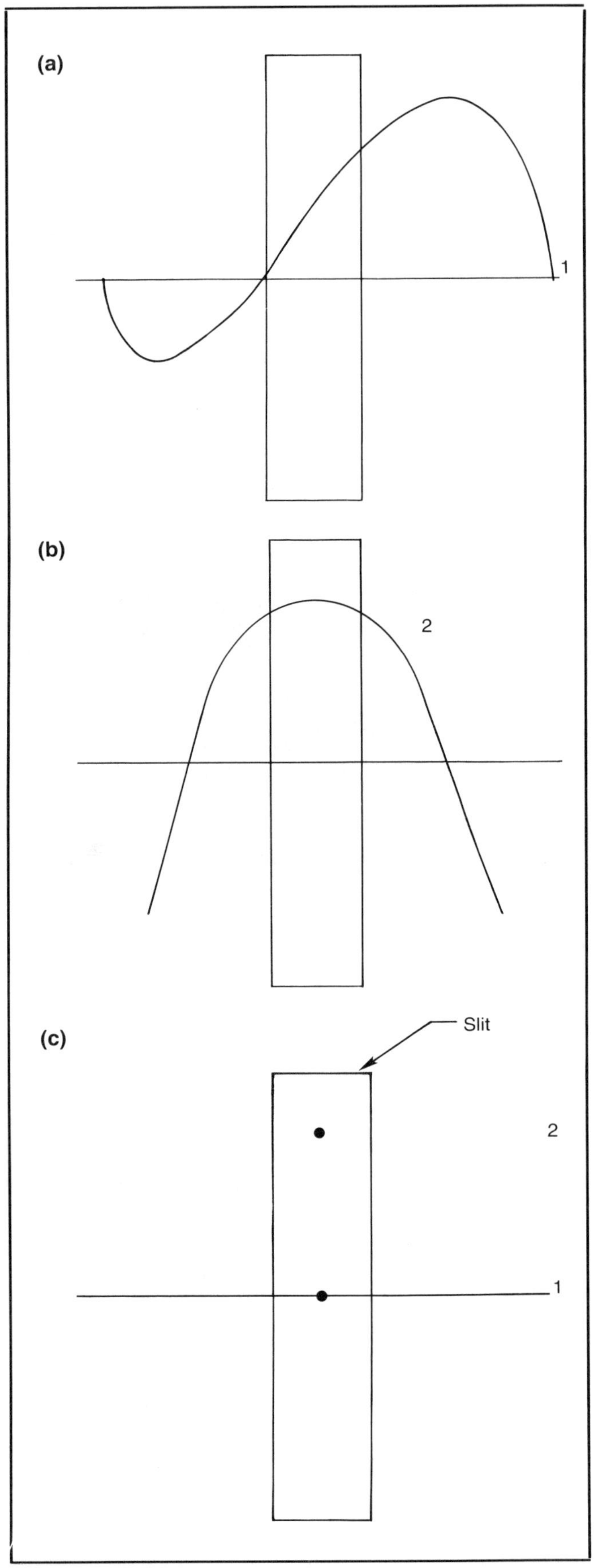

Figure 9-12. Time-slit method screen image of linear time-base instrument, with sinusoidal signals. (a) Heat-treated. (b) Annealed. (c) Combination.

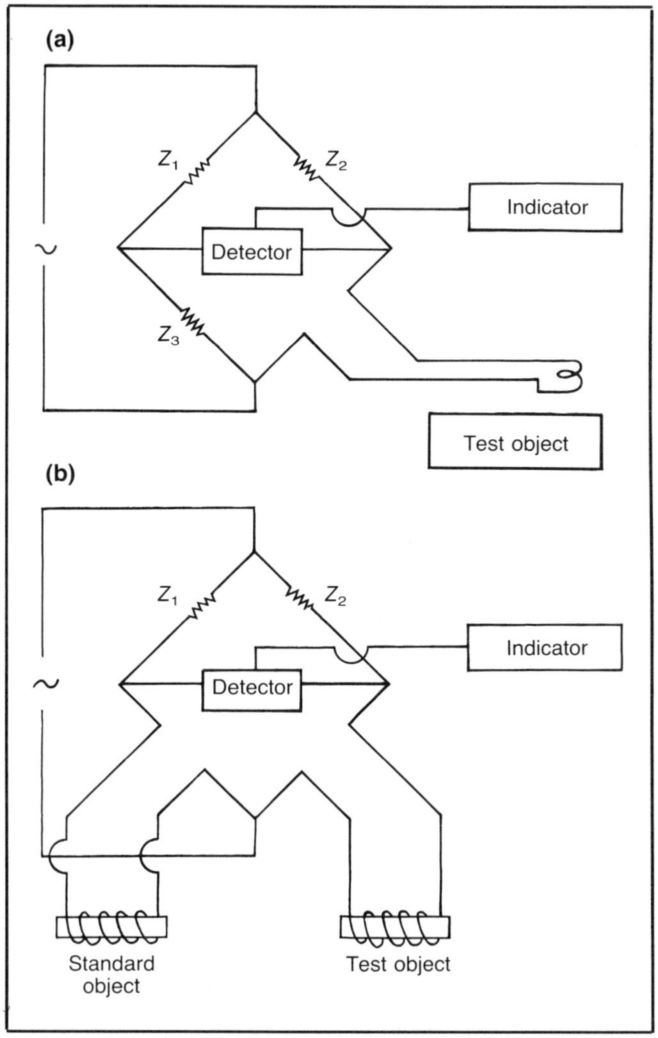

Figure 9-13. Test-coil bridge circuits. (a) Single test coil. (b) Two-coil comparison test.

and held manually. In this case, the demands for test-object handling equipment are minimal, and all that is required is a place to set or hold the test object while it is being inspected. In many tests, mechanical feeders feed the test objects past the test-coil assemblies so that tests can be made rapidly under uniform conditions. Such equipment requires coordination in the design of the test coil and the feed equipment.

Test instrument circuits vary greatly, depending upon applications. A common bridge circuit is shown in Fig. 9-13(a). The arms of the bridge are indicated by impedance symbols Z_1, Z_2, Z_3, and the test coil. Balance is obtained by adjustment of one or more of the bridge arms. The bridge can be balanced for a normal, acceptable test object, and the indicator is nulled to read zero. A variation in the test object will cause a change in impedance, resulting in a change in potential and hence a response on the indicator. The bridge may be made up of four resistance arms instead of the impedance arms shown. In either case, the ratio of inductive reactance (or resistance) of one pair of adjacent arms must equal the ratio of inductive reactance of the other pair of adjacent arms.

Another variation of this circuit is shown in Fig. 9-13(b). Here, another test coil has been added in an adjacent arm of the bridge, resulting in a comparison arrangement. In this case, standard test objects are first placed in each of the two coils and the bridge is balanced. Then, one of the test objects is replaced by an unknown test object, and the differences between the unknown test object and the standard are indicated by the degree of bridge unbalance. This bridge circuit can be used with more-involved amplifier phase detectors or sampling analysis systems.

In other types of detectors, the signal from the coil is broken down into its reactive and resistive components, and this information is either analyzed or presented directly. With equipment of this type, it is possible to separate wanted and unwanted signals because of the angle the two signals make with each other in the impedance plane. However, this type of analysis is useful only when there is sufficient angular difference between the curves for the two conditions to be separated. Systems of this type are sometimes referred to as phase-discriminating because of their ability to separate variables on the basis of the phase-angle differences in the impedance plane.

A feature found in many instruments today is liftoff compensation. As is evident from the description of the control, its function is to reduce sensitivity to liftoff so that the only signals that appear on the instrument are related to conditions of interest. In its commonest form, liftoff compensation is a means for varying frequency over a small range to locate the frequency at which sensitivity to liftoff is at a minimum. In any frequency range, there is a frequency for a given probe material-conductivity combination at which sensitivity to liftoff is a minimum. No liftoff compensation scheme gives infinite compensation; most systems can compensate only up to about 0.25 mm (0.010 in.) or a little more.

Instruments designed for scanning material—especially tube-testing equipment—will contain circuitry for differentiating the signal. When a signal is changing in a coil, it will change at some characteristic rate that is a function of scanning speed and the nature of the signal source. For instance, a change in dimension will normally occur rather slowly relative to the change measured for a crack. A differentiating circuit is an electronic means for separating the rapidly changing signals from the slowly changing signals: the slowly changing signals are blocked by the circuit, the rapidly changing ones are passed through. In this way an additional bit of information can be extracted from the inspection. This can be done safely if the slower-changing signals can be assumed to be of no interest in the inspection. This is generally the case for such things as dimensional variations in tubing.

For many applications it is desirable to automate decision-making to accept or reject a part on the basis of test results. For this reason, many instruments include some sort of alarm circuit. These circuits provide an output whenever a signal exceeds a preselected amplitude. Once such a signal is available, it can be used to activate any kind of auxiliary equipment. Lights, audible alarms, markers to identify the bad material, or systems to shunt the bad part into a separate bin or area can be operated by this alarm signal.

One type of instrument that is available allows the separation of any two variables by making it possible to select any arbitrary operating point within the impedance plane (see Fig. 10-2). In equipment with a meter, the resistive (*R*) and inductive (*X*) components in the coil impedance can be balanced separately. The circuit diagram is shown in Fig. 9-14. Once the characteristic curves for the material variables of interest are determined by plotting *X* and *R* balance points for various samples, the *X* and *R* balance controls can be adjusted to any desired point in the impedance plane. The instrument meter reads a voltage proportional to the distance from the chosen operating point to the curves. An operating point can generally be chosen near the center of curvature for the variable curve that is not of interest. This makes the signal amplitude resulting from small changes in this variable nearly constant. An example is shown in Fig. 9-15. As indicated, the two curves are for material thickness and liftoff. The operating point in this case has been chosen to minimize changes in meter reading related to liftoff while still giving a large variation in meter reading related to changes in thickness.

Looking at both the impedance change and the phase shift reveals much that is otherwise hidden in many testing situations. Therefore, conditions related to changes in conductivity can be considered separately from those relating to other influences such as permeability, liftoff, and dimensional, or thickness, changes. This advancement, known as impedance-plane analysis, has been a milestone in enhancing the detectability, interpretation, and, in turn, the reliability of the eddy current response information.

The brief discussion above describes some of the many differences in eddy current instrumentation. There are probably more different eddy current instruments on the market than instruments for any other nondestructive test method. As examples, there are instruments designed primarily for measuring electrical conductivity, inspecting tubes, inspecting wire, measuring coating thickness, and measuring magnetic properties. The multiplicity of test instrumentation for eddy current inspection results from the variety of test parameters that are required for this type of test. To build a single, all-purpose eddy current instrument would be impractical because each test situation requires a slightly different set of test parameters. To cover a large number of situations would make the instrument much too complex and expensive. Hence, there are many different types of instruments available, each intended to cover only a limited amount of the vast field of eddy current applications.

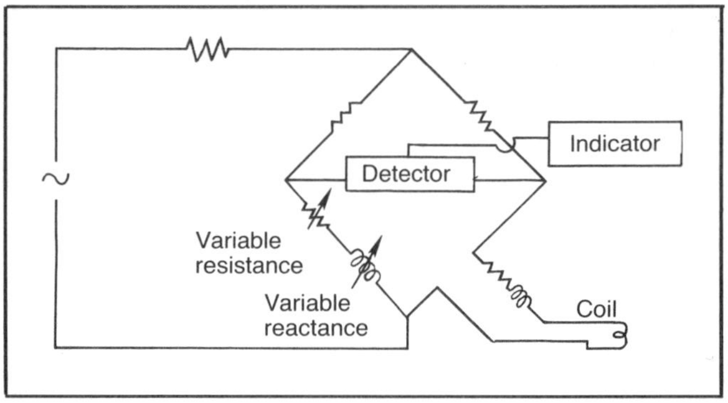

Figure 9-14. Test-coil bridge circuit for variable resistance and reactance.

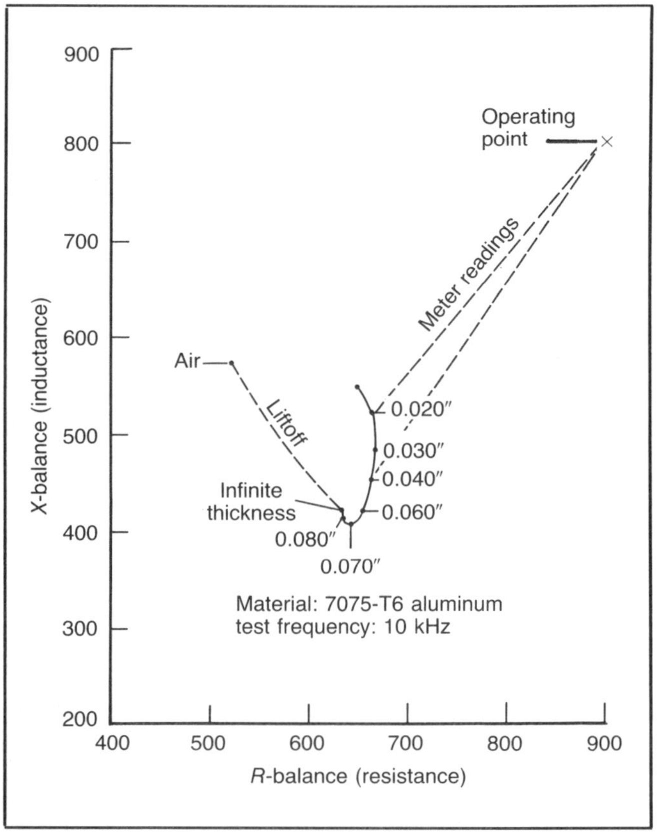

Figure 9-15. Impedance-plane diagram showing selection of an operating point that maximizes thickness and minimizes liftoff in the instrument readout.

Reference Standards

Introduction

Most eddy current tests are qualitative in nature in that the test results of accept/reject decision criteria are based on some measure of comparison to a reference standard. Without these standards, it is difficult to calibrate or standardize the instrument to obtain consistent results from day to day. If the instruments are not consistently calibrated, then the inspection results are meaningless. Additionally, no comparison can be made between different inspection agencies or personnel when testing the same object or product. Therefore, instrument calibration or reference standards are mandatory to establish accept/reject decisions and maintain systems reliability. Reference standards are also required for those eddy current tests that yield quantitative values, such as conductivity meters and thickness gages. Here, known value reference standards are used to calibrate the instruments before performing the test or inspection.

Reference standards may be divided into the following types, or categories:

1. Thickness testing.
2. Alloy sorting (metallurgical changes).
3. Crack detection and depth measurement.
4. Dimensional measurement (diameter, wall thickness, spacing, etc.).

Thickness-Testing Standards

Thickness-testing types, methods, and applications are discussed in chapter 10. The three main types of eddy current thickness tests are (1) metal foil, (2) nonconductive coating on conductive base, and (3) conductive coating on a conductive base.

Regardless of the type of test, what is needed is one bare specimen and three others: one of minimum thickness, one of maximum thickness, and one in the mid-range of thickness. The instrument is calibrated using the above standards and then the unknown-thickness specimen is measured.

Alloy Sorting (Metallurgical Variations)

There are times when two or more alloys are inadvertently mixed because of error of substitution. These error mixes may result in some parts having improper mechanical or physical properties, which may result in improper performance or part failure. The discovery of the improper mix usually follows a discrepancy report followed by a laboratory evaluation. What is known is that the material should be made of alloy A when in fact it is made of alloy B.

The first thing that is needed for the eddy current test evaluation is known standards of alloy A and alloy B. With these in hand, the technician can establish a sorting test that can be applied by an inexperienced person in a "go"-"no-go" fashion. In certain cases, the parts may be made from the right alloy but some are improperly heat-treated. All that is needed in this case is two standards—one that is properly heat-treated and one that is improperly heat-treated.

Although the eddy current test is measuring differences in conductivity or resistivity, what is being sorted is parts having differences in hardness and strength. A relationship can be established between hardness, conductivity, and strength for a particular aluminum alloy in a particular temper. Generally, as the strength decreases, so do the hardness values. However, the conductivity increases as both hardness and strength decrease (at a particular temper). The change (increase) in conductivity with a decrease in strength for 7075 alloy in the T6 temper is shown in Fig. 9-16. If absolute values of strength are to be obtained by eddy current methods alone, the time and temperature of the overaging must be known and constant.

When standards are selected, the following conditions and precautions should be taken into account:

1. The standards must be of the same alloy and heat-treatment condition as the parts to be inspected.
2. The standards should have the same size and shape as the parts to be inspected (see Figs. 9-17 and 9-18).
3. No variations in plating, coatings, or surface finish are allowed between the standards and the parts to be inspected.
4. Long parts such as rods or tubes should be long enough that no end effect occurs.
5. Do not cold-work the standards by bending or rubbing.
6. Do not overheat standards.

The above conditions are valid for both ferrous and nonferrous material standards.

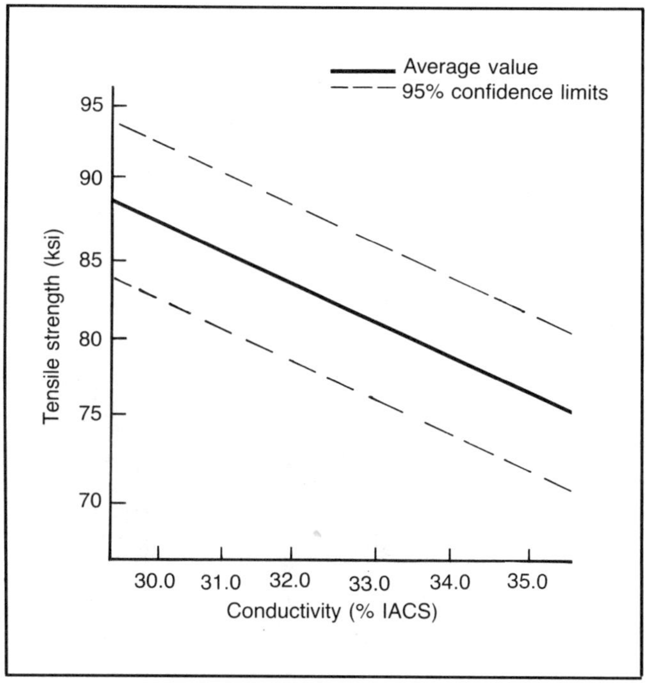

Figure 9-16. General relation between tensile strength and eddy current conductivity measurements of 7075-T6 aluminum (FM-110).

When eddy current instruments are used to determine physical properties of ferromagnetic materials, the measurements depend upon the magnetic permeability μ. This type of inspection is generally referred to as a magneto-inductive test.

The basic principle of a magneto-inductive test is directly related to the principles of eddy current testing. In most equipment used for testing ferromagnetic materials, the part is placed in encircling coils. When this is done, energy losses occur. This is the result of a combination of eddy current and magnetic characteristics. These losses are called "core" or "power" losses. When measured with the proper equipment, they can be distinguished from each other. The relationship between these electromagnetic properties and the physical properties is not as clearly established for ferromagnetic materials as it is for nonmagnetic materials.

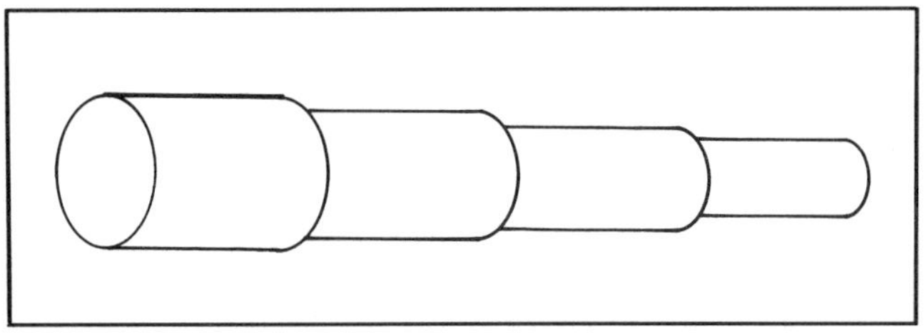

Figure 9-17. Variable-diameter test standard.

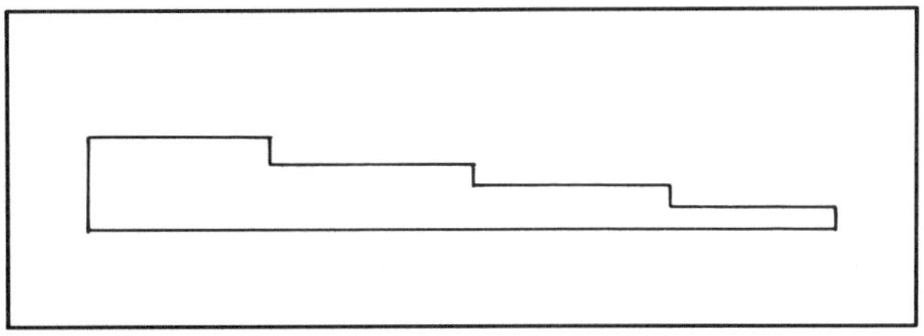

Figure 9-18. Variable-thickness and specimen-width test standard.

A correlation between the properties does exist. However, there are a number of possibilities that should be taken into account when ferromagnetic materials are involved. The power losses do not vary directly with the hardness (strength) throughout the full range of hardnesses. In the pearlitic microstructure, a large variation in hardness will show only a fairly small variation in the power losses. Through the upper and lower bainite structures, a reversal occurs, and the power losses start decreasing. When martensite is the major microstructure, a small change in hardness (strength) will give a proportionately large change in power losses. This is shown graphically in Fig. 9-19 and is referred to as the electromagnetic curve. It should be noted that in the area of reversal, there is also a loss in ductility.

When hardness (strength) measurements are made, it is necessary to plot the curve to determine the validity of the test. This demands a number of standards of the various hardnesses. The standard Rockwell test can be performed to determine the hardness of the standards. The indentation caused by this test does not interfere with the subsequent

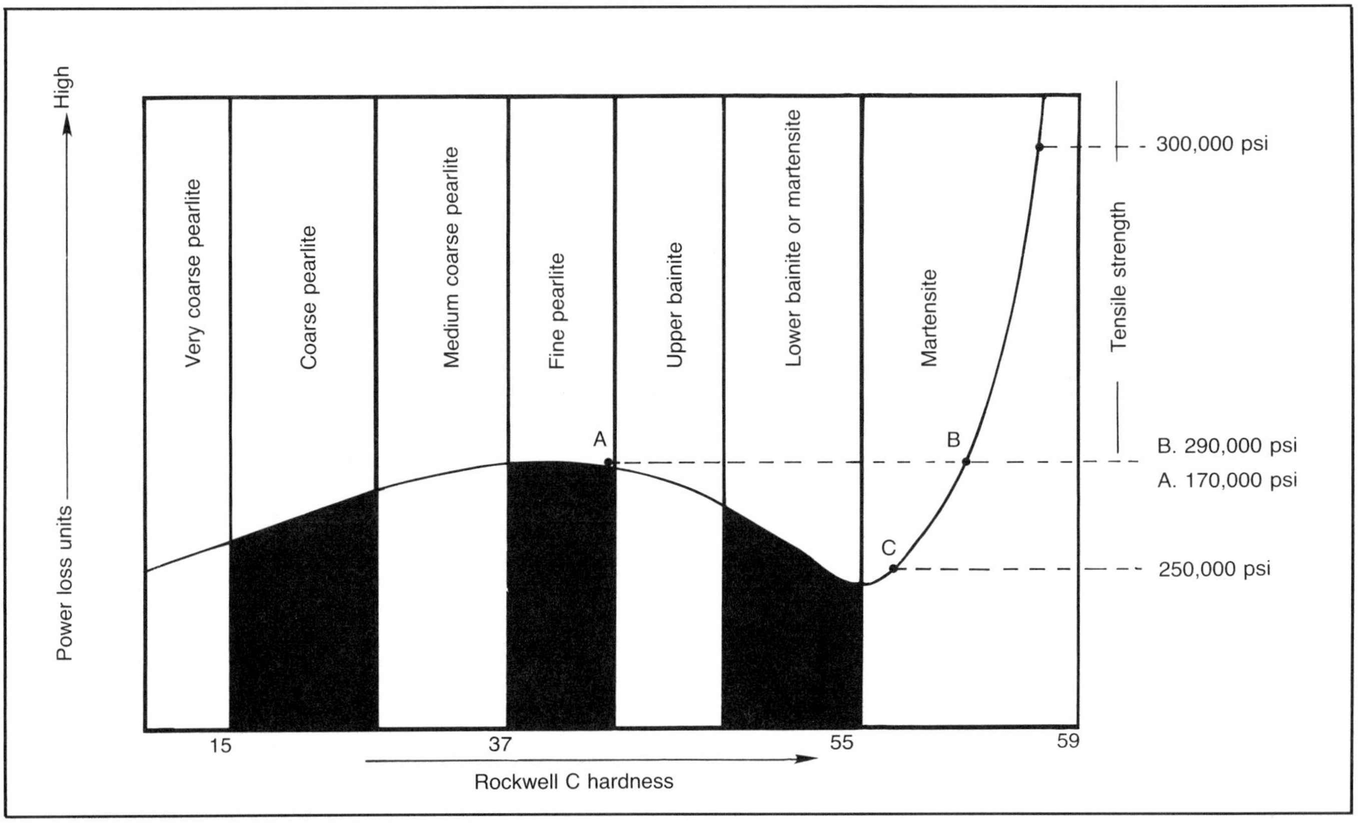

Figure 9-19. Electromagnetic S curve for medium carbon steel.

magneto-inductive test. In the martensitic range, magneto-inductive measurements within ±1 R_C are well possible.

When these tests are carried out, it is essential that the alloy is known and is the same for all parts that are to be tested. When alloys have to be separated, it is necessary that the hardness of the parts be the same. The best possibility for either test is the use of a low-frequency (15 to 400 Hz) instrument using two encircling coils in which a known standard is compared with the unknown parts under inspection. A probe-type instrument can also be used, though the frequency will generally be over 300 Hz to keep the probe to a reasonable size.

Measurements at higher frequencies, over 60 kHz, can also be made. In that case, the skin effect makes the influence of the electrical resistance more significant. The use of an encircling coil is recommended because a more stable reading can be obtained than with a probe. A probe can be used only when a smooth inspection surface is available. For added stability, the probe should be spring-loaded.

The relevant data obtained with the instruments mentioned here are displayed on either an oscilloscope or a simple meter readout. The oscilloscope generally displays a linear time base with one cycle for viewing. Because the time base can be shifted in phase, it is possible to distinguish between permeability and conductivity variation because the latter is perpendicular (normal) to the permeability variations. The phase angle for diameter variations is practically the same as for permeability variations, so the dimensions of the parts under inspection have to be very close to constant to ensure a reliable hardness test.

It should be noted also that the parts are to be placed in or moved through the coils in the same position to prevent erroneous readings. This is especially important when the process is automated. The instruments using a meter readout will give only the magnitude of the variations; in many instances, this is quite sufficient. In automated sorting systems, whether for nonferrous or ferromagnetic materials, the parts are passed through encircling coils. This can be done by either gravity or a transport system. The latter method is generally preferred for larger parts and for asymmetric parts. When low-frequency equipment is used, the speed of the inspection should be considered critically.

The eddy current information received from the part passing through the coil is analyzed by an electronic switching circuit, the output of which will trigger an alarm or mechanical sorter. Two- or three-way separations can be made in this manner. The various types of mechanical sorting devices used generally depend on the parts under inspection. To ensure a fail-proof inspection, a positive switching mode should be used, so that when equipment failure occurs, defective or uninspected parts cannot be marked acceptable.

Flaw Detection

Numerous eddy current tests are performed to detect such flaws as cracks, pits, and corrosion. Standards for flaw detection should simulate, as closely as possible, the condition being inspected for. In addition, the length, depth, and severity of the artificial or real flaws should represent conditions on which accept/reject decisions can be made by the eddy current inspector.

Fig. 9-20 shows a typical standard for use in calibrating and detecting cracks in aluminum aircraft structures. The three calibration slots for a surface probe are of different depths. Usually, these notches are 0.125 mm (0.005 in.) wide and 0.25, 0.5, and 1.25 mm (0.01, 0.02, and 0.05 in.) deep. The notches at the corners of the holes are also 0.125 mm (0.005 in.) wide by 0.75 mm (0.03 in.) deep. The instrument is calibrated to the appropriate notch before inspection, and cracks approximately equal to, or greater than, the notch depth will give a recognizable response on the instrument readout.

In nonferrous materials, the depth of penetration is fairly well known, and when the diameter of the probe coil exceeds the length of the flaw, the depth can be approximated. The width of the flaw is not known, so the accuracy of depth measurement is only an approximation. When the crack depth exceeds the depth of penetration, no measurable increase in eddy current response is obtained.

Standards with seams of varying depth, and destructive tests, are necessary to establish the relationship between the eddy current data and crack depth (see Fig. 9-21). In ferromagnetic materials, the depth of penetration is not easily determined because the magnetic permeability is the major influencing factor and is seldom accurately known.

Notches in the above standards are usually made by electrical-discharge machining (EDM) to obtain the desired notch width and depth.

Standards for pit detection are best made by drilling the desired-diameter flat-bottom holes at varying depths in the standard: 10, 35, 50, and 75

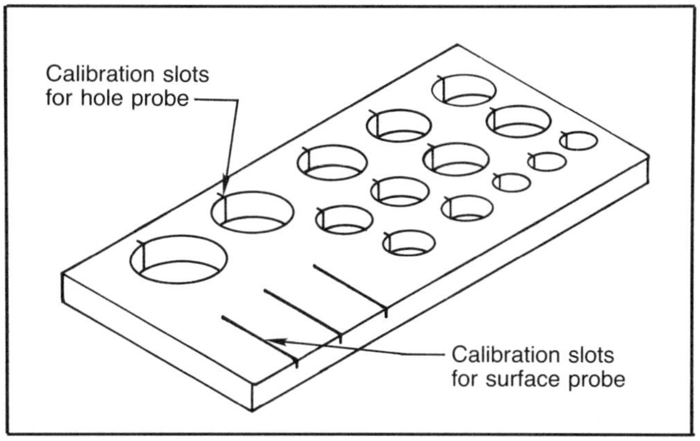

Figure 9-20. Calibration block for bolt-hole and surface-probe inspections.

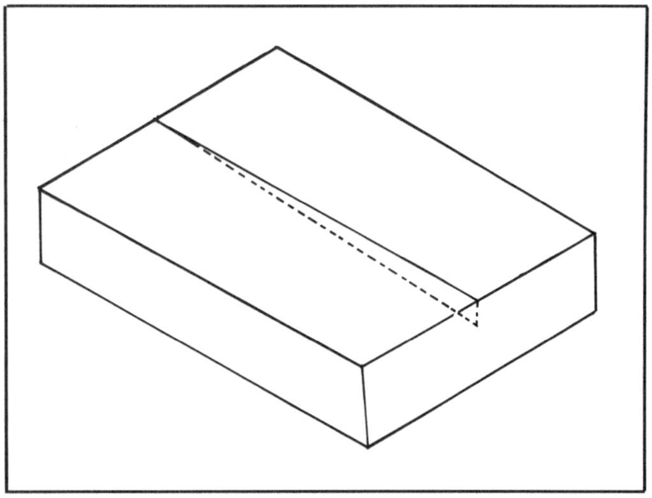

Figure 9-21. Eddy current standard with preselected depth.

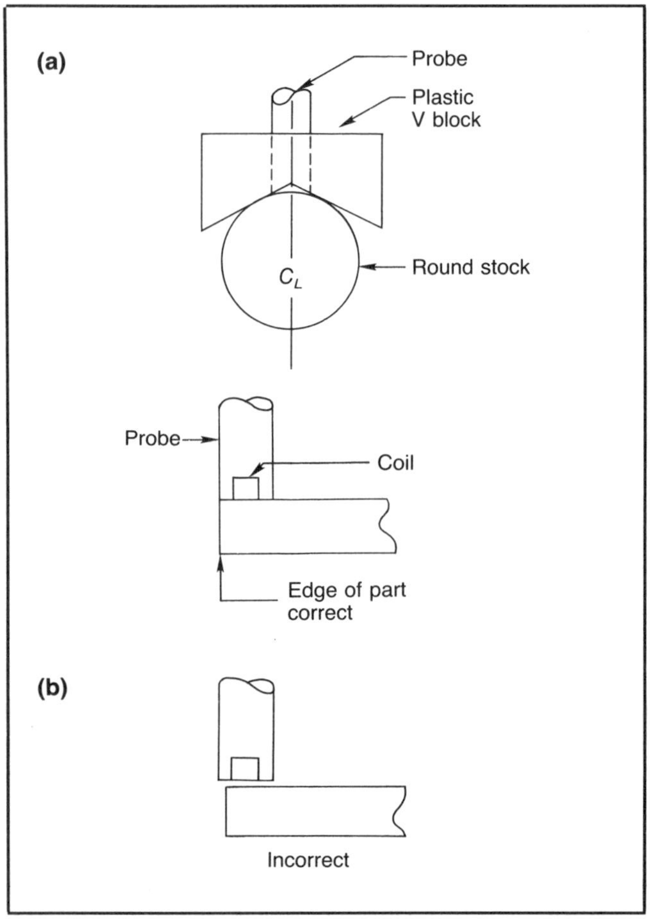

Figure 9-22. Ways to avoid edge effects or probe wobble (with probe coil and cylindrical parts). (a) End view, good coupling between test object and probe coil. (b) Side view, use of probe coil with cylindrical test object.

percent depth and, usually, one hole that penetrates through the standard. The instrument is calibrated to yield variable response from the drilled holes, and flaw depth can be approximated by

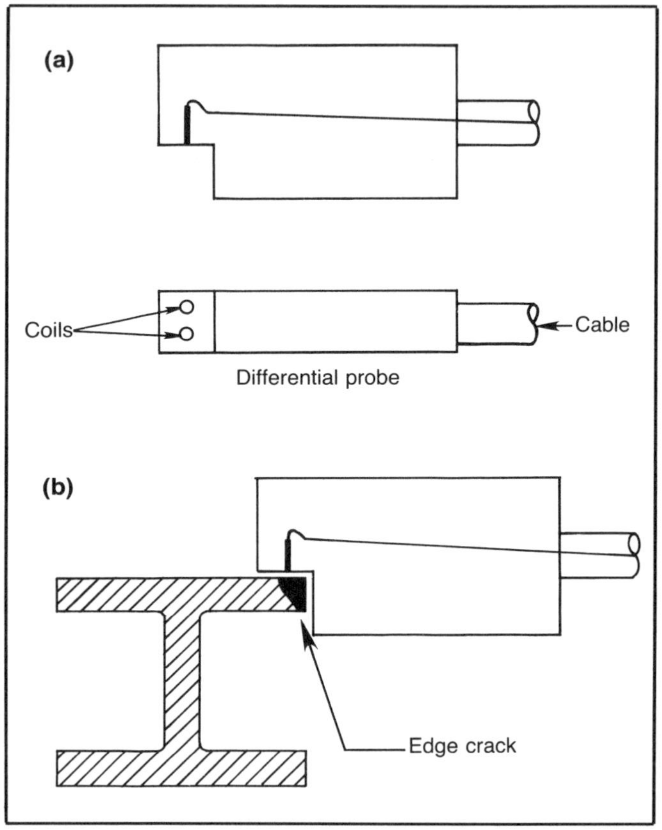

Figure 9-23. Edge-guide probe used to avoid edge effects during inspection for edge cracks. (a) Edge-guide probe. (b) Eddy current check for edge cracks.

comparing its response to that obtained from the reference holes.

Dimensional Measurement

When making eddy current dimensional measurements, all variables, except the one being measured, must be controlled and constant. Unwanted variables will influence the results and accuracy of the desired measurement. Hence, the reference standard must be similar to the parts being tested with respect to conductivity, permeability, and geometry.

If thickness changes are being measured, then the fill factor or liftoff and operating frequency must be kept constant. A step wedge similar to that shown in Fig. 9-18 is usually sufficient for instrument calibration. The steps in the wedge should be of the desired thickness so that accept/reject decisions can be made concerning the test parts. A similar-type standard is useful for determining thinning due to corrosion.

A typical standard representing changes in diameter is shown in Fig. 9-17. As the part diameter decreases, the inductive reactance also decreases. The result is similar to that obtained from increasing liftoff (see Fig. 10-7) or decreasing fill factor (see Fig. 8-1). Correction factors may be established for changes in diameter. To avoid probe wobble and also maintain alignment between the probe coil and axis of the part, a V guide should be used; see Fig. 9-22(a).

Narrow parts may be evaluated by placing the probe on the edge of the step standard shown in Fig. 9-18. As the wedge gets narrower, the inductive reactance also decreases, resulting in a response similar to that of liftoff. When testing narrow parts, a V guide should also be used to maintain alignment between the coil and the part centerline.

When an eddy current coil is brought close to the edge of a part, a very strong signal is obtained as a result of "edge effects;" see Fig. 9-22(b). Although the eddy current response close to an edge is different from response from areas away from the edge, it is still possible to inspect along the edges, provided the coil is kept a fixed and constant distance from the edge.

If cracks initiate at an edge of the part, then a reference standard having an edge notch is used to calibrate the instrument. A special probe or probe guide is required to maintain a fixed distance between the edge of the part and the eddy current coil. Such a probe is shown in Fig. 9-23(a). The inspection procedure for detecting edge cracks is illustrated in Fig. 9-23(b).

Typical reference standards are available for use in eddy current determination of spacing between two conductors. Changes in spacing will yield a typical impedance-plane phasor change, as shown in Fig. 10-13.

Chapter 10
Impedance-Plane Response

Introduction

Eddy current testing consists of measuring, in a test coil, impedance changes caused by eddy currents induced in an electrically conductive material. Because impedance is the product of resistance and reactance, which are perpendicular (normal) to each other in the electronic sense, impedance changes may be plotted on an *X-Y* graph for the various test conditions. This plot is called the *impedance-plane diagram.* These diagrams are two-dimensional displays of both amplitude and phase of the test response.

Older eddy current instruments reflected impedance changes by meter deflection. Newer phase-analysis instruments enable the operator to produce impedance-plane phasor responses automatically on the integral *X-Y* storage oscilloscope. If the test samples show a significant difference, test feasibility is immediately established. These instruments operate from 60 Hz to 6 MHz, allowing the operator to choose the best frequency for a given material and test. Different material conditions, such as conductivity, permeability, cracks, liftoff, spacing, and thinning, produce their own unique impedance-plane plots. It is this capability that allows rapid, significant impedance-plane analysis in a wide variety of applications.

Looking at both the impedance change and the phase shift reveals much that is otherwise hidden in many testing situations. Therefore, conditions related to changes in conductivity can be considered separately from those related to other influences such as permeability, liftoff, and dimensional, or thickness, changes. This advancement, known as impedance-plane analysis, has been a milestone in enhancing the detectability, interpretation, and, in turn, the reliability of the eddy current response information.

It is common practice to plot the reactance X_L as the ordinate and resistance R as the abscissa in the impedance plane. In this way, the test-coil impedance Z is represented by a point P formed by two perpendicular (normal) components, X_L and R, on the impedance plane. In the absence of a test object, the empty test coil has a characteristic impedance, with coordinates X_{L_0} and R_0, shown on the impedance plane by the coil in air (the "air point"), point P_0 of Fig. 10-1. If the probe is placed on the test object, the original field of the coil in air is modified by the superimposed field of the eddy currents. This field modification has exactly the same effect as would be obtained if the characteristics of the test coil itself had been changed. The influence of the test object can be described by a

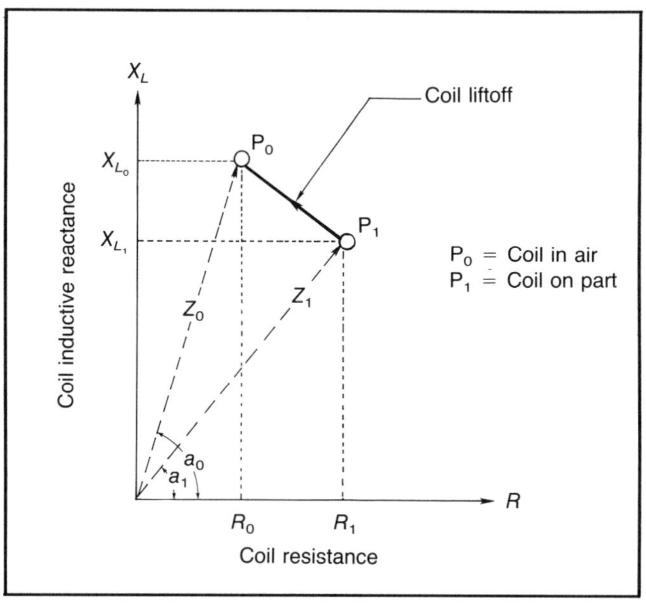

Figure 10-1. Representation of the test-coil characteristics on the impedance plane.

variation in the test-coil characteristics. The apparent impedance of the "coil in air" represented by P_0 is displaced to P_1 (corresponding to new values of X_L and R) under the influence of the test object (see Fig. 10-1).

The magnitude and direction of the displacement of the apparent impedance from P_0 to P_1 under the influence of the test object are functions of the properties of the test object and the characteristics of the instrumentation. Significant properties of the test object include

1. electrical conductivity (σ) or resistivity (ρ),
2. dimensions of the test object,
3. magnetic permeability (μ), and
4. presence of flaws such as cracks.

Significant instrument characteristics include

1. frequency (f) of the alternating current (AC) field in the test coil,
2. size and shape of the test coil, and
3. distance of the test coil from the test object (liftoff or fill factor).

Real values of impedance and phase angle may be calculated from complex equations, which may then be normalized to establish changes on the impedance plane caused by either material or instrument characteristics (see Figs. 7-2 and 7-3). In a practical testing situation, using a cathode-ray tube (CRT) flying-dot impedance-plane instrument, the operator tries to simulate previously published normalized impedance diagrams.

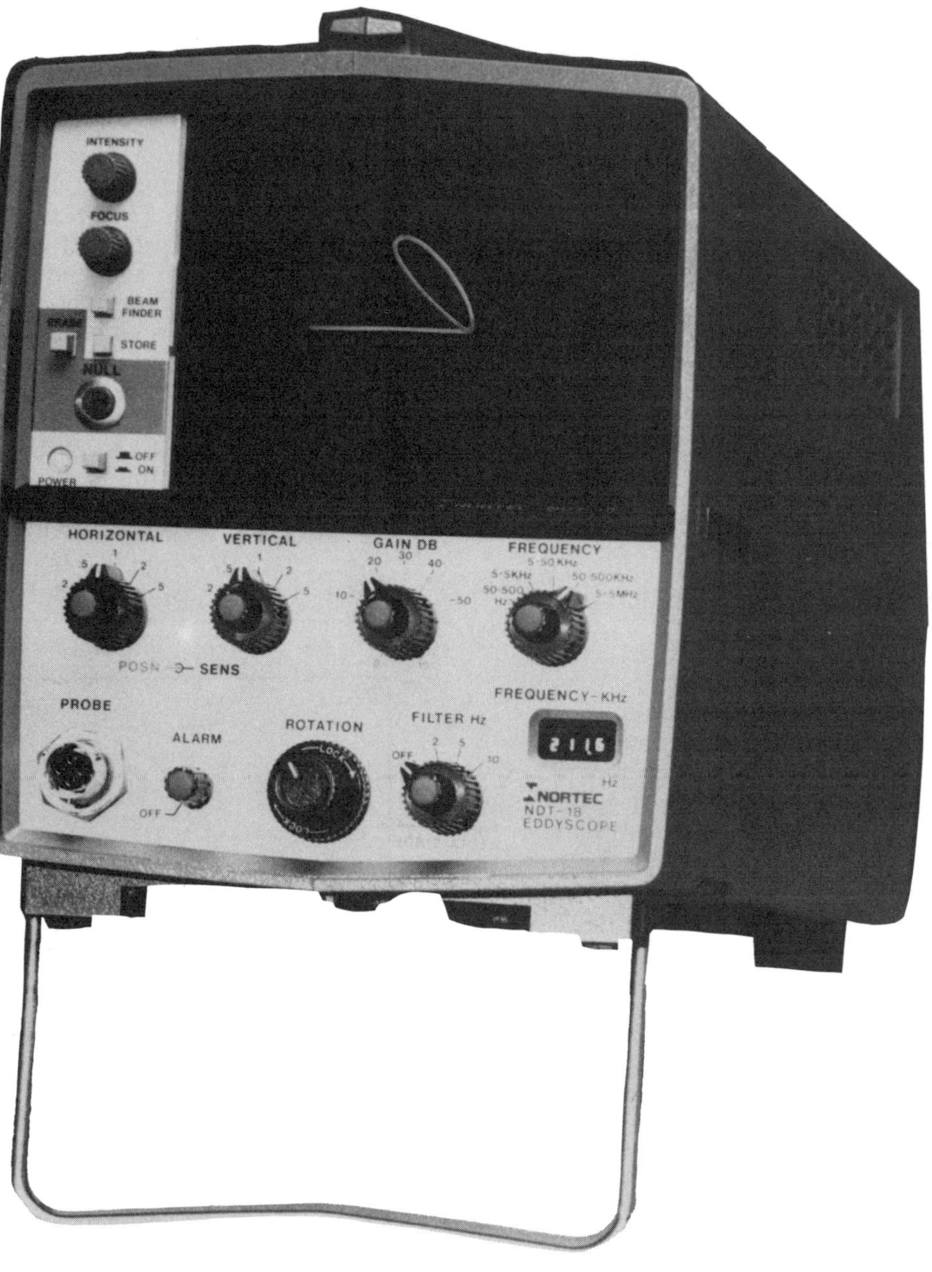

Figure 10-2. Typical CRT "flying-dot" impedance-plane instrument.

Memory Oscilloscope Presentation of Impedance Changes

A typical CRT flying-dot impedance-plane instrument is illustrated in Fig. 10-2. After connecting the test coil (probe) to the instrument and setting the frequency to that printed on the test probe, the instrument is calibrated in accordance with manufacturer's instructions.

To produce typical *phasors* for changes in liftoff, conductivity, and permeability, flat specimens of different metals should be available. These specimens should be about 50 mm (2 in.) square and have a thickness greater than 3δ. When the probe is placed on successive specimens and the instrument controls are properly set, a phasor diagram similar to Fig. 10-3 should be obtained. This display shows how the coil's impedance is affected by liftoff, conductivity, and permeability changes.

If we join the *impedance vector* (phasor) points with an imaginary line, two comma-shaped curves are produced; the upper one represents changes in magnetic permeability (μ) and the lower one represents changes in conductivity (σ) or resistivity (ρ), as shown in Fig. 10-4. Note that changes in *permeability* and *resistivity* trace out the comma curve in a counterclockwise direction, whereas increasing *conductivity* changes in a clockwise direction.

Fig. 10-5 shows the effect of test frequency on the conductivity and liftoff curves for nonmagnetic alloys. As can be seen, changes in frequency shift the points along the conductivity locus in a nonlinear fashion. This phenomenon, also true for other impedance curves, can be advantageously used because it allows the material points to be located for optimum response/suppression. Specifically, a frequency should be chosen that causes the material points for the variable to be measured to move in a substantially different direction from those points to be suppressed.

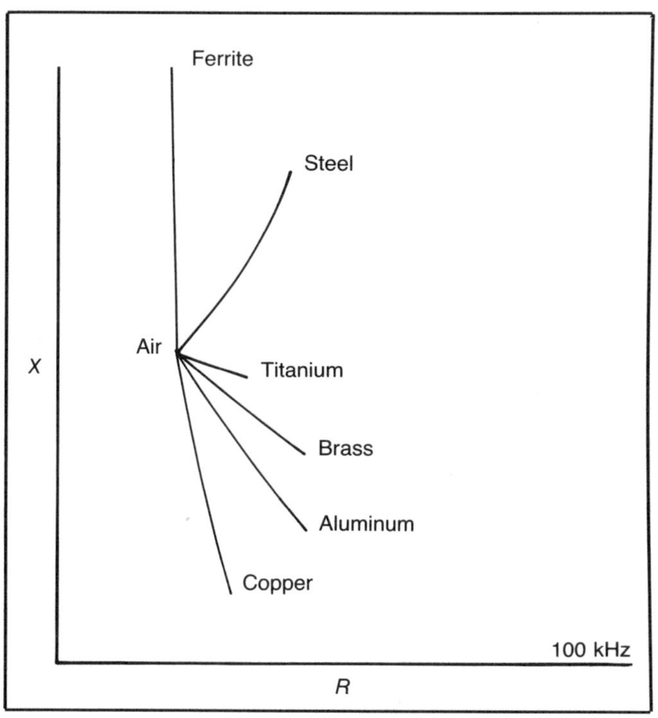

Figure 10-3. CRT impedance-plane presentation for different metals.

Liftoff and Edge Effects on the Impedance Plane

The probe-to-part spacing is referred to as *liftoff*. Fig. 10-6 shows the impedance-plane response that occurs when the spacing is increased between the probe and part. The upper portion of the impedance plane is the magnetic domain, where responses occur from ferromagnetic materials. The lower portion is where responses are obtained from nonmagnetic materials (after proper calibration).

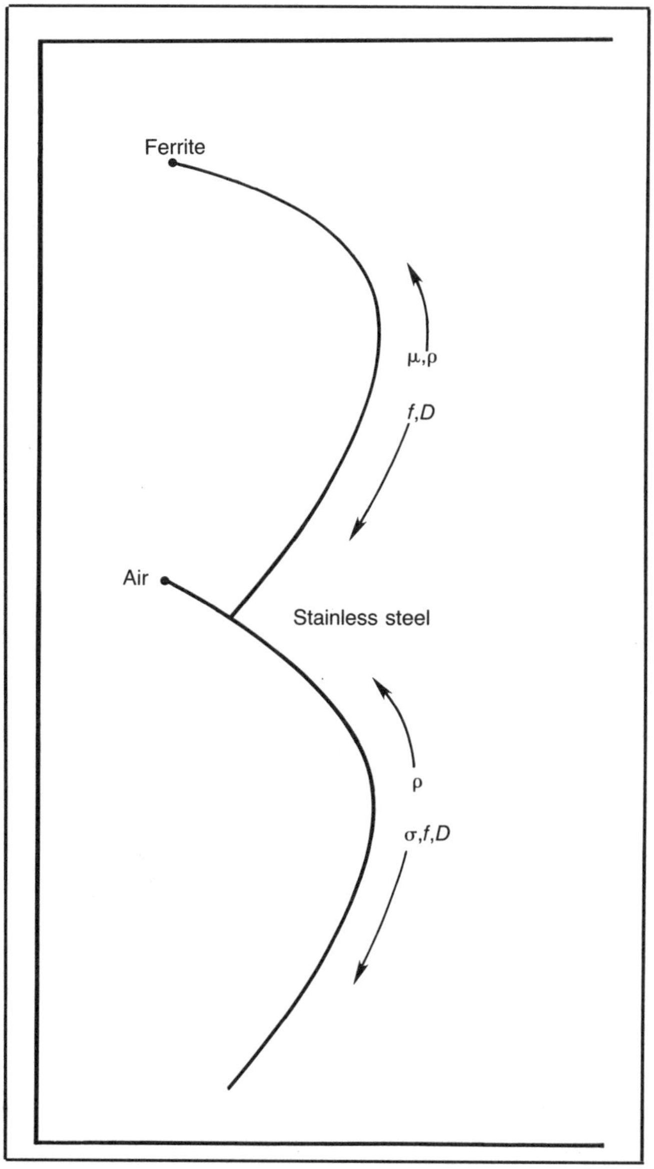

Figure 10-4. Locus of points producing the typical comma curves for permeability (μ) and conductivity (σ) or resistivity (ρ) using a surface probe of diameter *D* operating at a frequency *f*.

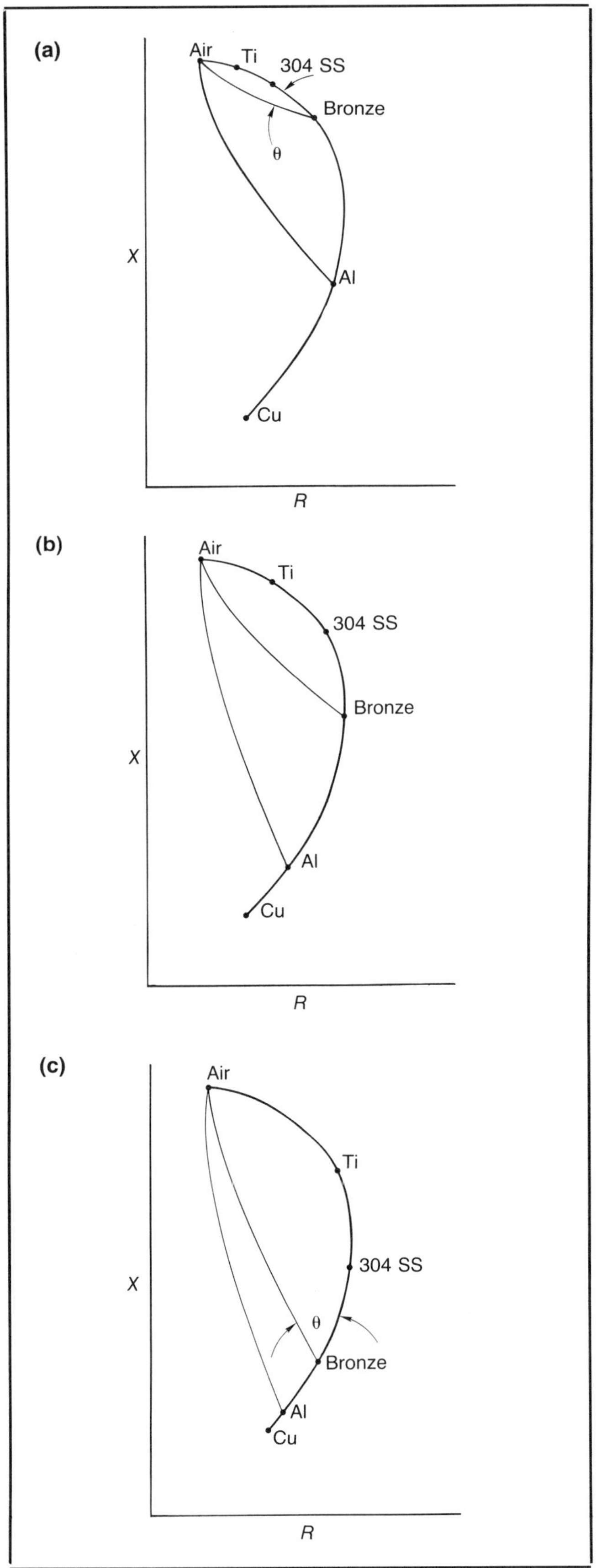

Figure 10-5. Movement of material points by frequency changes. (a) Low frequency (20 kHz). (b) Medium frequency (100 kHz). (c) High frequency (1 MHz).

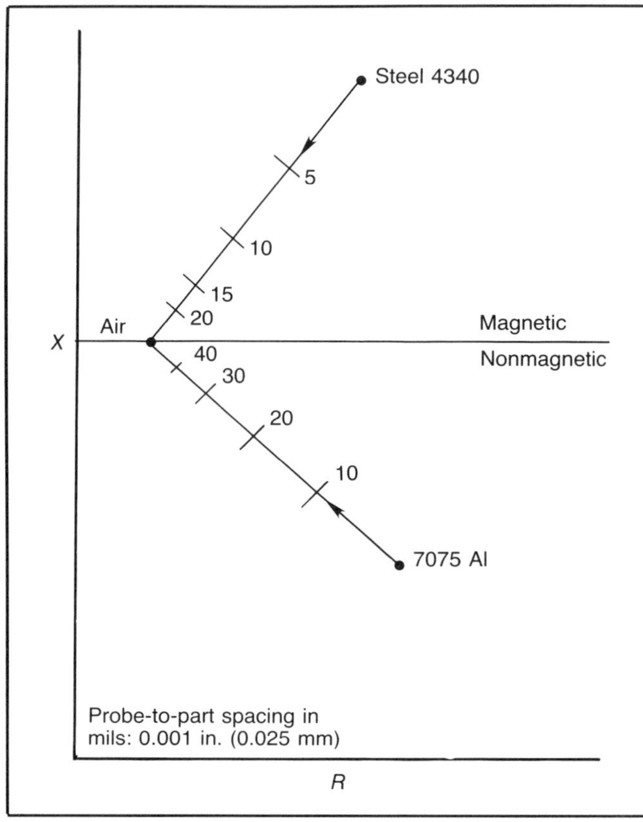

Figure 10-6. Liftoff effect on impedance plane at 100 kHz.

Note the nonlinear (logarithmic) changes along the liftoff locus for equal increments of spacing.

When the probe is moved near the edge of the specimen or part, an *edge effect* occurs because a portion of the magnetic field is outside the part. For nonmagnetic materials, the result is a response similar to the liftoff response. The liftoff and the edge effect from an aluminum sample are shown in Fig. 10-7(a) at 10 kHz, 100 kHz, and 1 MHz. For ferromagnetic materials, the spacing between liftoff and edge-effect loci is slightly greater than for nonmagnetic materials, possibly because of a greater distortion of the magnetic field; see Fig. 10-7(a). The example shown in Fig. 10-7(b) was performed at 10 kHz, 100 kHz, and 3 MHz.

Conductivity and Permeability Loci on the Impedance Plane

The impedance-plane response to different *conductivities* of various *nonmagnetic* alloys is shown in Fig. 10-5. The material points trace out a characteristic comma curve, with conductivity increasing in a clockwise direction. The coil (probe)-liftoff loci are shown for bronze and 7075-T6 aluminum. Note that the separation angle (θ) between the conductivity locus and the liftoff locus (for a given material) is much smaller for bronze when it was tested at 20 kHz than when it was tested at 1 MHz. Hence, the unwanted liftoff variable will affect test results less when low-conductivity materials are tested at a high

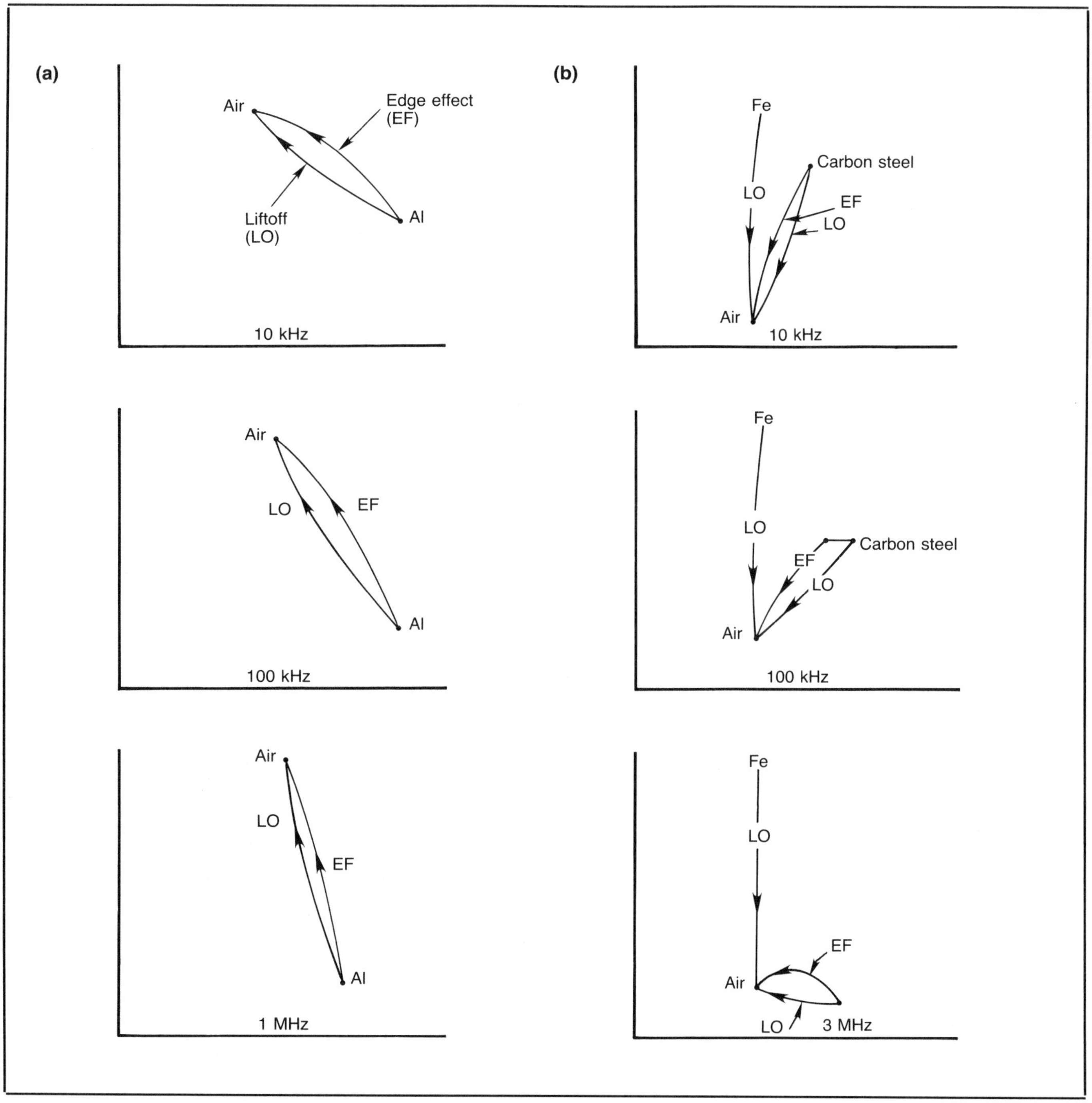

Figure 10-7. Impedance phasor diagrams for liftoff and edge effects at different frequencies. (a) Nonmagnetic materials. (b) Magnetic materials.

frequency. Fig. 10-5 also shows the effect of test frequency on the conductivity and liftoff curves for nonmagnetic alloys. As can be seen, changes in frequency cause the points along the conductivity locus to shift in a nonlinear fashion.

When ferromagnetic materials are placed in or near the magnetizing-coil field, the magnetizing-coil reactance changes in a far different way than with nonmagnetic test materials, as previously shown in Fig. 7-3 and Fig. 10-4. When a high-permeability test material is used, the magnetizing-coil inductance and inductive reactance increase dramatically because of an increase in flux density. When testing with surface probes, an increase in permeability moves the operating point up the impedance locus, as shown in Figs. 10-3 and 10-4. A surface probe with a ferrite-core coil yields better magnetic coupling and hence yields a larger impedance diagram than a probe with a similar air-core coil. Also, magnetic permeability has the same effect as resistivity (ρ), and therefore these two parameters usually cannot be separated when a surface probe is used (see Fig. 10-4).

Fig. 10-8 shows the complex impedance plane for the case in which ferromagnetic test bars of various relative permeabilities entirely fill the

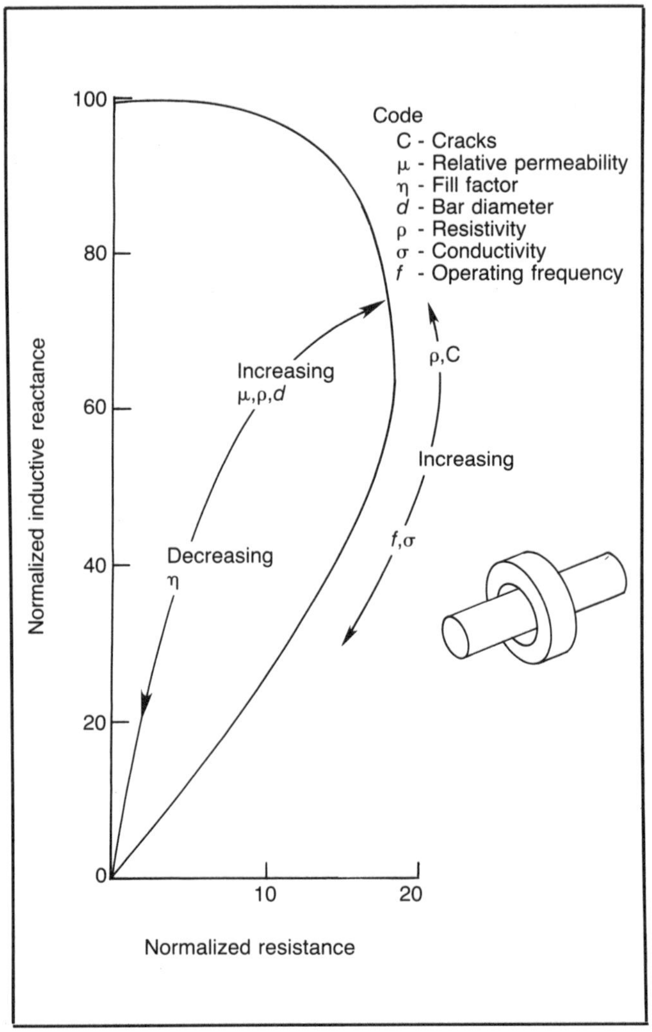

Figure 10-8. Normalized impedance graph for ferromagnetic bars using an encircling coil. Note the high values for normalized inductive reactance and resistance. Changes in relative permeability and bar diameter lie along the same locus and cannot be separated.

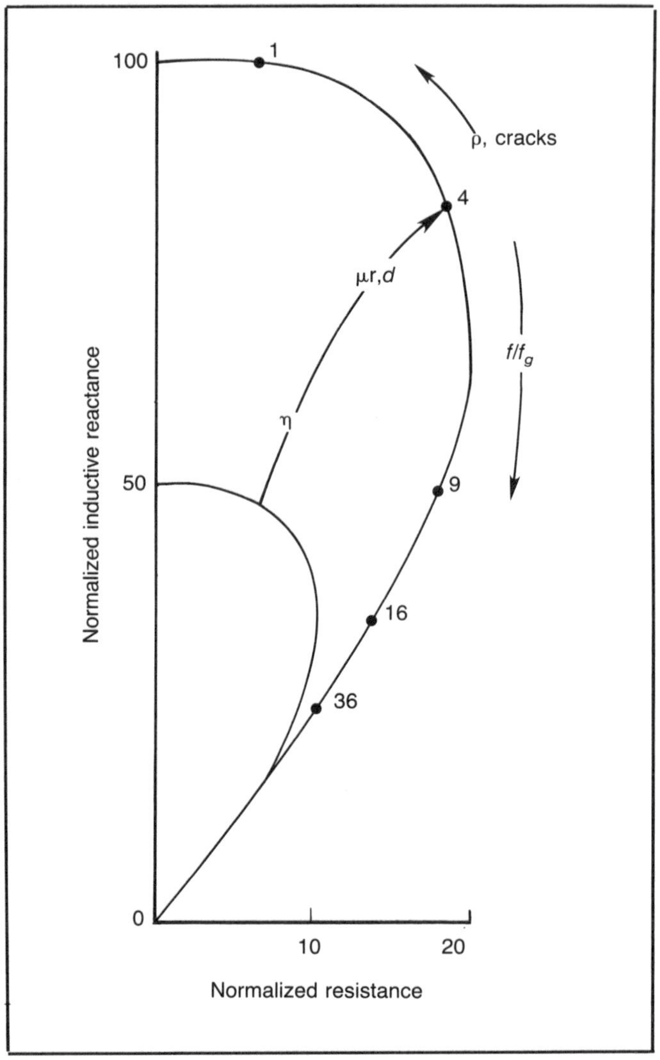

Figure 10-9. Normalized impedance graph for ferromagnetic bars tested with an encircling-coil system and fill factor of unity. Note that there is good separation between cracks and permeability/bar diameter for f/f_g values less than 4.

annulus of an encircling-coil test system. This set of curves differs from the preceding complex-plane conductivity locus curves for nonmagnetic bars because the magnitudes of both the real and imaginary components of the coil impedance with ferromagnetic test bars are increased by a factor equal to the relative magnetic permeability of the high-permeability test material. With a test bar of steel with a relative magnetic permeability of 100, the magnitudes of encircling-coil test signals could be as large as 100 times those resulting from nonmagnetic test bars. The horizontal and vertical scales of the complex-plane diagram must thus be increased by the same factor to accommodate the enlarged signal-locus curves for high-permeability test materials.

Fortunately, during testing of steel bars for surface cracks, the common direction of diameter and permeability changes on the complex plane of test signals is at a relatively large angle to the direction of changes caused by cracks. Thus, the interfering effects of variations in bar diameter and magnetic permeability can be separated (by phase analysis) from the indications of surface cracks. Magnetic permeability variations can arise from such conditions as internal stresses resulting from bar straightening. Such variations are often very annoying difficulties in eddy current tests of steel parts with encircling test-coil systems. In many cases, direct current (DC) saturating coils are used to reduce these effects when the purpose of the encircling-coil tests is flaw detection rather than sorting by diameter or material properties.

In the complex plane of impedance coil voltage signals for ferromagnetic bars (see Fig. 10-8), the direction of signal point displacement with changes in the bar diameter is identical to that of changes in the effective permeability. Both of these effects result in movement of the signal point along the locus curves of Fig. 10-8. Separation of effects of conductivity changes from the effects of changes in the diameter or relative permeability of ferromagnetic bars is most feasible in the upper half of the complex plane (Fig. 10-9). In the test-frequency range for f/f_g ratios (where f_g is the limit frequency) smaller than 4, changes in conductivity (which are a function of alloy content and of ferromagnetic-material structures related to hardness) can be measured independently of the effects of changes in relative magnetic permeability (due to mechanical stress following cold-working operations such as drawing or straightening). These measurements of alloy content can also be made independent of the

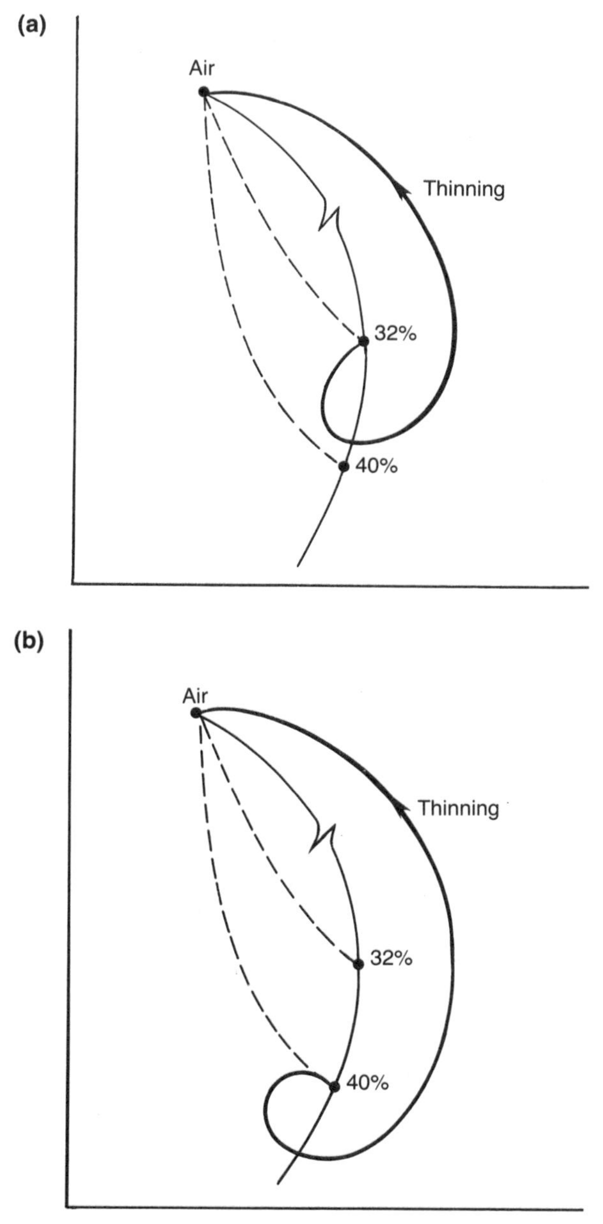

Figure 10-10. Impedance-plane thickness plots at 60 kHz. (a) 7075-T6 taper gage 32% IACS (5.4 $\mu\Omega$ – cm). (b) 7075-T73 taper gage 40% IACS (4.4 $\mu\Omega$ – cm).

effects of the diameter variations always present in commercial bars.

In the absence of a test bar, the empty-coil point P_0 for encircling-coil tests is at the same locus point of the complex-plane diagram for either nonmagnetic or ferromagnetic test bars. The point P_0 has a vertical signal component of unity and a horizontal signal component of zero. On the expanded scale required for high-permeability ferromagnetic materials, this point lies on the vertical axis and very close to the origin (0,0). When a ferromagnetic test bar is placed within the encircling test coil, the signal point moves from this point near the origin to a point on the locus curve for a given value of relative magnetic permeability, which is determined by the test-frequency ratio f/f_g.

When the eddy current instrumentation displays the test points on a CRT screen in a reproduction of the complex plane, the empty-coil point may be so close to the origin that the insertion of the test bar appears to move the beam spot from the origin to the proper coordinates on the complex-plane display. In many instruments, the portion of the complex plane near the origin is suppressed, and the enlarged image of a small portion of the complex plane is displayed on the face of the CRT. In this case, when the ferromagnetic test bar is removed from the encircling test-coil system, the empty-coil point lies far offscreen and cannot be observed.

Special difficulties arise in encircling-coil tests of ferromagnetic bars when variations can occur both in *bar diameter* and in the *relative permeability* of the bar material. These two test-object properties act together in the product factor that appears in all the significant equations for encircling-coil test signals from ferromagnetic bars. Both of these factors act along the same signal displacement directions in the complex-plane diagram constructed in Fig. 10-9 for *reduced coil fill factors* (η) and ferromagnetic bars. For example, a reduction in bar diameter and/or in its relative magnetic permeability can result in a reduction of the test-signal amplitudes. Thus, it is impossible to distinguish, by encircling-coil tests made at a single test frequency, whether the signal change is caused by a variation in bar diameter or by a change in relative magnetic permeability. When magnetic saturation techniques are used, the effects of changes in relative magnetic permeability are reduced or eliminated.

Metal-Thickness Variations

The strength of eddy currents decreases rapidly beneath the metal surface, as illustrated in Fig. 7-5. Penetration decreases with increasing frequency, increasing electrical conductivity, and increasing magnetic permeability. These three variables are used to define the standard depth of penetration, as previously defined in Chapter 7.

The magnetic permeability is 1.0 for nonmagnetic materials, and the conductivity of a particular material is generally known and constant. Hence, the depth of penetration is controlled by the operating frequency.

To compare the theoretical and experimental values for the standard depth of penetration and impedance-plane phasor movements caused by *metal-thickness changes*, one may take a taper gage and scan it at different frequencies. In the following example, two aluminum taper gages were scanned at 60 kHz. The 7075-T6 gage had a resistivity of 5.4 μΩ-cm (32% IACS), and the 7075-T73 gage had a resistivity of 4.4 μΩ-cm (40% IACS). The probe was scanned from the thick end of the taper gage toward the thin end, and phasor movement was plotted as shown in Fig. 10-10.

Fig. 10-11 shows the general effect on the coil impedance caused by changes in conductivity, thickness, and liftoff. The curved solid line represents the locus of specimen conductivity values. The spiral curve represents the locus of specimen thickness values, and the dashed lines represent the locus of liftoff values. The separation angle is defined as the angle, on the impedance plane, between the locus of liftoff changes and the locus of metal thinning. This angle increases with plate thickness and becomes important when impedance analysis is used to minimize the effects of liftoff.

If it is desired, for example, to measure changes in specimen thickness while incurring minimum interference from liftoff, the separation angle should be chosen near 90 degrees (see Fig. 10-12). However, if it is desired to measure conductivity or conductivity-related conditions throughout a metal sheet while minimizing interferences from thickness changes, the separation angle should be chosen near 180 degrees (see angle *B* in Fig. 10-12). Thus, the effects of thickness changes occur in the same direction as liftoff, and both effects can be separated simultaneously from conductivity changes.

A closer look at Fig. 10-11 reveals that although δ is an accurate and mathematically defined depth of penetration, it can, if used, lead to inaccurate test results. In this figure, the conductivity values obtained at δ thickness are far removed from the conductivity locus plot. To obtain accurate conductivity results, choose a thickness value that looks infinitely thick to the eddy currents. This thickness value is approximately equal to 3δ.

However, if one is trying to measure changes in thickness while minimizing liftoff, the test frequency should be chosen so that the separation

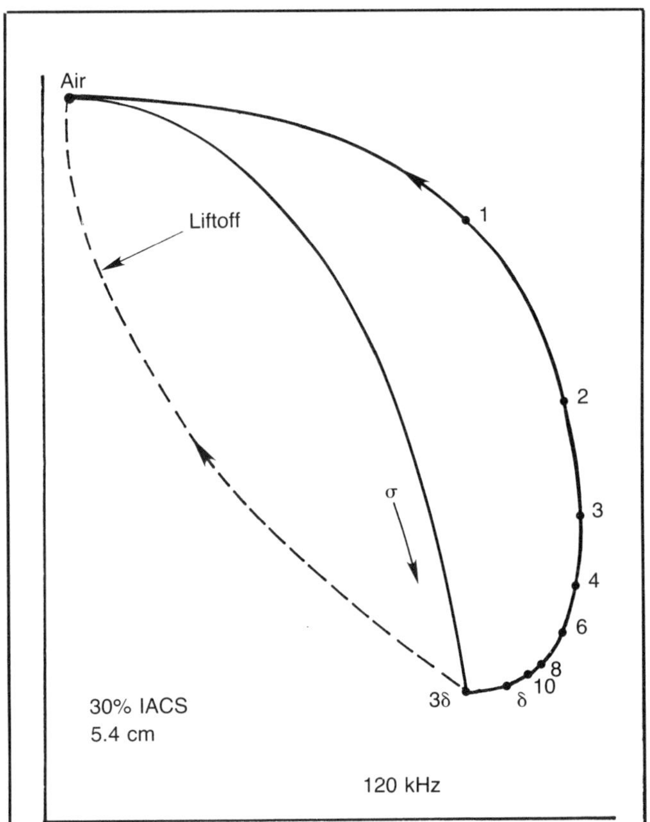

Figure 10-11. Typical impedance-plane diagram for liftoff, conductivity, and thickness from aluminum at 120 kHz (probe coil). 0.001 in. = 0.025 mm.

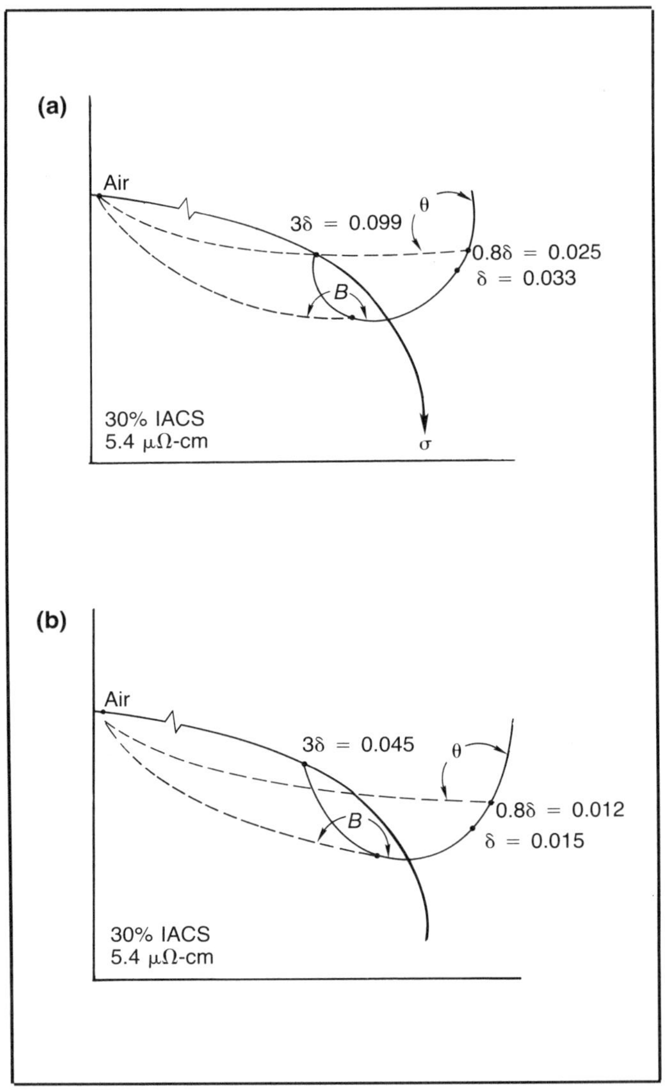

Figure 10-12. Apparent-thickness impedance diagram for aluminum at (a) 20 kHz and (b) 100 kHz. 0.1 in. = 2.5 mm.

angle between them is 90 degrees. This frequency can be calculated using the depth-of-penetration formula (see Chapter 7). A reasonable approximation for thin sections is obtained when

$$t = 0.8\delta \qquad \text{(Eq. 10-1)}$$

which converts to

$$f = 1.6\ \rho/t^2\text{, in kHz} \qquad \text{(Eq. 10-2)}$$

where

δ is skin depth in mm,
t is material thickness in mm, and
ρ is material resistivity in $\mu\Omega$-cm.

Fig. 10-10 shows the overall relationship of thinning for two different-conductivity specimens at a selected test frequency (60 kHz); Fig. 10-12 shows an expanded view of the thinning response from a 5.4 $\mu\Omega$-cm (30% IACS) specimen tested at 20 kHz and 100 kHz. At a value of 3δ, the specimen looks infinitely thick to the eddy currents; hence, the point lies on the conductivity locus. As the specimen gets thinner, the vector point spirals downward in a counterclockwise direction until the thickness reaches a value of about 2δ, where it crosses the conductivity locus and continues in an upward direction toward the air point (zero thickness). The 3δ, δ, and 0.8δ (determined by Eq. 10-1) values are illustrated in Fig. 10-12(a) for 20 kHz and 10-12(b) for 100 kHz. Although the 0.8δ values define a separation angle of 90 degrees between the liftoff and thickness loci, for practical thickness testing one may start with δ thickness and work downward through thinner values, as shown in Fig. 10-11. Also in this figure, note that the spacing between the phasor points increases as the thickness decreases.

Plating Thickness

Fig. 10-6 shows the liftoff effect when a magnetic or nonmagnetic material is coated with a nonconductive material. However, a vastly different impedance-plane plot is obtained when a conductive coating is applied to a conductive base material. The layered examples shown in Fig. 10-13 do not necessarily represent a practical situation, but rather have resulted from matching some arbitrary combinations of metals to illustrate how these combinations behave on the impedance plane. However, the principles illustrated do have practical applications. The impedance point of thick copper-base material would be a point on the conductivity locus, as shown in Fig. 10-13(a).

Now, if the copper-base material is plated with aluminum, the probe-coil impedance moves upward and to the left (clockwise) of the conductivity locus. The impedance approaches the point (3δ) on the conductivity locus corresponding to thick aluminum. If copper of various thicknesses is added on the top of aluminum, the probe-coil impedance follows a locus downward and to the right of the conductivity locus. The movement of the impedance point (with increasing thickness) is not proportional to the thickness;

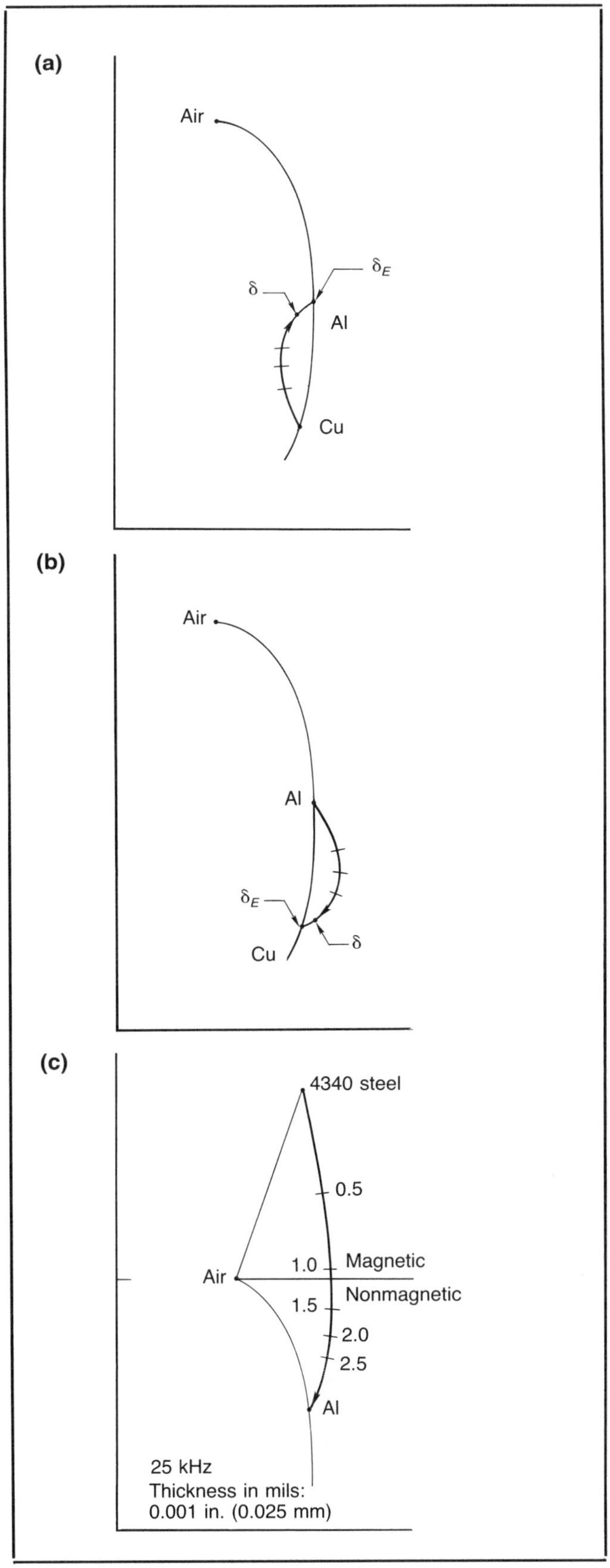

Figure 10-13. Impedance graphs—coating thickness effects. (a) Low-conductivity coating on high-conductivity material. (b) High-conductivity coating on low-conductivity material. (c) Aluminum coating on steel.

equal thickness changes become less as the thickness increases. The impedance point corresponding to a thickness approximately equal to one standard depth of penetration δ is shown in Fig. 10-13(b). The impedance point finally converges quite rapidly on the copper conductivity locus at 3δ.

The important fact illustrated in Figs. 10-13(a) and (b) is that the curves spiral with increasing curvature in a clockwise direction. The start of the spiral in each case approximates a straight line but as the thickness increases the spiral becomes more accentuated. The degree of spiraling and the shape of the curve will vary as the operating frequency is changed.

Finally, we have the case of nonmagnetic coating on a magnetic substrate or base material, as illustrated for aluminum on steel in Fig. 10-13(c). In this example, the inductive reactance decreases in a nonlinear fashion. The locus of points is to the right of the permeability and liftoff locus line. The locus crosses from the magnetic domain to the nonmagnetic domain on the impedance plane at rather thin values of aluminum. Similar results would be obtained for cadmium (7 μΩ-cm) and chromium (21 μΩ-cm) plating on steel alloys.

Metal Spacing

There are times when a gap separates two metal sheets. The gap may be filled with a nonconductive adhesive layer or a nonmetallic shim or purposely planned as a fixed dimension. If it is desired to measure the gap or spacing, an operating frequency and coil diameter must be chosen so that eddy currents will be generated in the second (subsurface) layer. The frequency and probe diameter chosen should produce eddy currents that penetrate the upper layer to a depth approximately equal to δ and both metals only to a depth of 3δ (see Chapter 7).

When the gap is zero (both metals touching), the combination appears infinitely thick and produces a vector point on the conductivity locus. As the gap is increased, the inductive reactance increases, and the phasors produce a locus of points in the vertical direction which cross the conductivity locus at some point. As the gap increases, the locus of points will finally intersect the thinning locus at a point that represents the thickness of the upper metal specimen (see Fig. 10-14). At this point, the gap has reached a value where eddy currents are being generated only in the upper layer. Hence, any further increase in the gap between the two metal specimens will not cause any further change in impedance. If a larger gap needs to be measured, the operating frequency must be reduced and/or a larger coil diameter must be selected.

Gap or spacing measurements are best made with the probe located over the thin member, which reduces the depth-of-penetration problem. If both members are of equal thickness but one has a lower conductivity, the probe should be placed on the lower-conductivity specimen.

Cracks

Eddy current inspection is used frequently to find cracks in aircraft structures during in-service maintenance checks. The use of impedance-plane analysis has provided a better understanding of crack response loci with other variables on the impedance plane. This is evident in Fig. 10-15(a), where the crack response locus is 90 degrees out of phase with the permeability (μ) and liftoff (LO) locus for 4340 steel. Because of the wide separation angle between these two loci, cracks in steel are fairly easy to detect. The lower amplitudes and phase angles for electrical-discharge machining (EDM) notches in the steel sample are also shown in Fig. 10-15(a).

Fig. 10-15(b) shows that for aluminum alloys, the EDM notch and crack response lies between the liftoff and conductivity loci. The smaller notch response almost parallels the liftoff response, but the angle of separation and amplitude increase as the notch depth increases. A fairly good response is obtained from a deep crack in a similar aluminum specimen. It can be seen in the figure that the crack response is almost parallel to the conductivity locus. This characteristic is true for all nonmagnetic and magnetic materials, as illustrated in Fig. 10-16.

A deep surface crack in nonmagnetic material looks like a reduction in conductivity, and therefore a frequency that puts the material near the knee of the conductivity curve would give best crack sensitivity. Higher frequencies push the material point down the curve, push conductivity points closer together, and therefore reduce sensitivity to a given conductivity change (crack). Fig. 10-17 confirms the previous rationale. In this figure, the conductivity locus points for specimens having conductivity values in % IACS of 1, 3, 8, 20, 30, and 40 are shown. The crack is in the 30% IACS specimen. At 10 kHz, the conductivity points are well spaced along the conductivity curve, and the

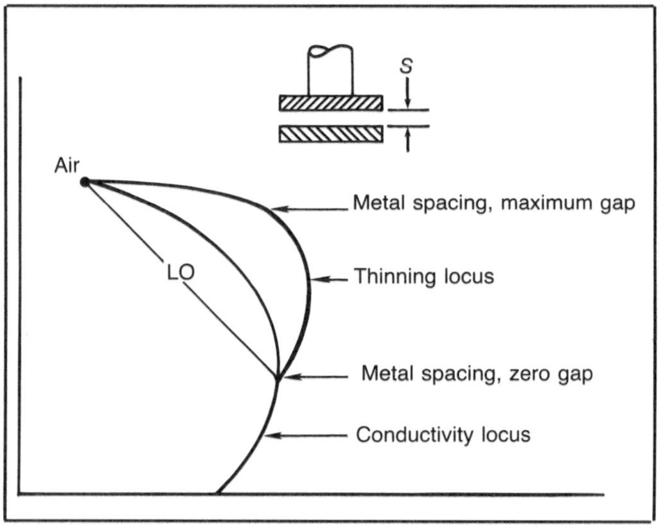

Figure 10-14. Metal spacing impedance-plane diagram.

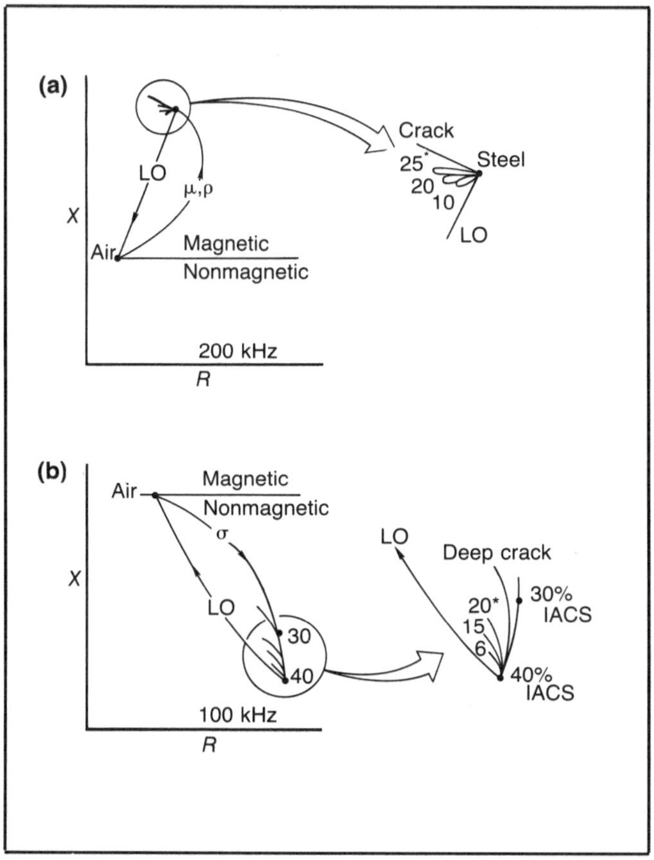

Figure 10-15. Crack/notch responses on impedance plane. (a) EDM notches and crack in steel. (b) EDM notches and crack in aluminum. *EDM notch depth in mils: 0.001 (0.025).

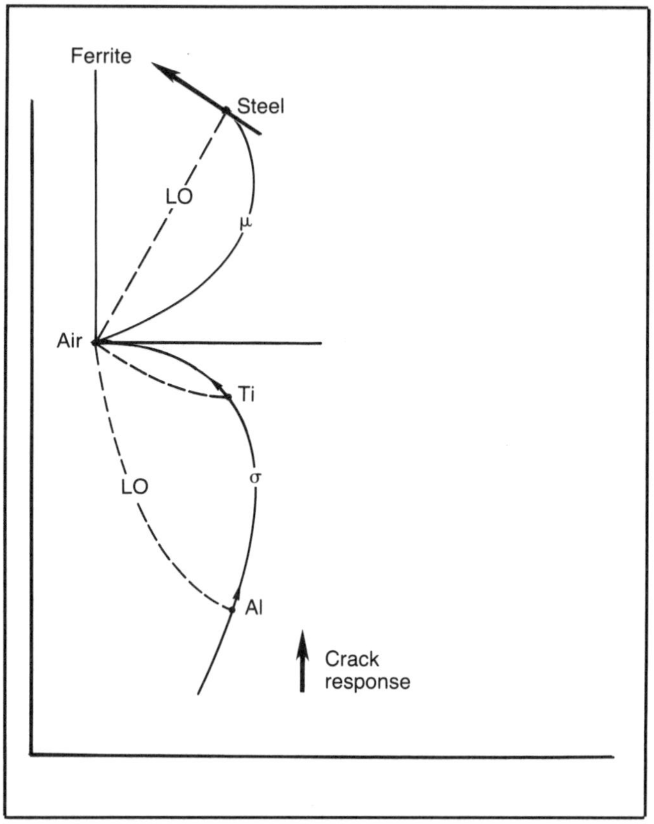

Figure 10-16. Direction of surface cracks on the impedance plane.

separation angle between the liftoff and crack is about 45 degrees. As the frequency is increased, the crack signal amplitude drops, the material points move closer together, and the separation angle between the liftoff and crack gets slightly smaller. Compare Figs. 10-17(a) at 10 kHz with Figs. 10-17(b) and (c) at 100 kHz and 3 MHz, respectively.

Generally, deep cracks (greater than 3δ) interrupt most of the eddy current flow and look like a bulk effect (reduced conductivity), causing an increase in reactance. However, as the crack gets shallower, it looks more and more like liftoff, and the limiting case of the zero-depth crack has zero amplitude and a phase angle in the direction of liftoff.

If the crack is at least 2 to 3δ deep at a frequency that puts the material point near the knee of the conductivity curve, that is the best frequency. Shallow cracks require a frequency for which crack depth equals 2 to 3δ, but any higher frequency would reduce response.

Fig. 10-18 shows impedance phasor diagrams for a crack in 4340 steel at 10 kHz, 100 kHz, and 3 MHz. Here, the liftoff response to ferrite remains vertical because of the low resistivity. As the frequency is increased, the steel point moves down the impedance plane, and the angle between the liftoff and crack does not change much.

Note how good crack detection looks at 3 MHz. Liftoff is a small locus, indicating low sensitivity to small changes, and the crack is large and at a right angle to liftoff. Deep cracks look like a local reduction in conductivity, whether in nonferromagnetic or ferromagnetic materials (see Fig. 10-16).

In the early 1970s, engineers at Boeing and Douglas aircraft companies investigated subsurface crack detection in aircraft structure using low-frequency eddy current inspection techniques. Fig. 10-19 shows how through-the-thickness cracks in second- and third-layer aluminum structures appear on the impedance plane at 1 kHz. The surface-crack response in the first layer runs almost parallel to the liftoff locus. The second-layer crack runs parallel to the conductivity locus, and the third-layer crack response is to the right of the conductivity locus.

Impedance Changes in Relation to One Another

Previous figures showed individual changes on the impedance plane for various conditions such as conductivity, permeability, cracks, metal spacing, and metal thinning. These examples were for both magnetic and nonmagnetic materials. Fig. 10-20 illustrates the relationships (position, shape, and phase angle) of the various phasor plots on the impedance plane. The figure data may be used to determine how each test problem reacts on the

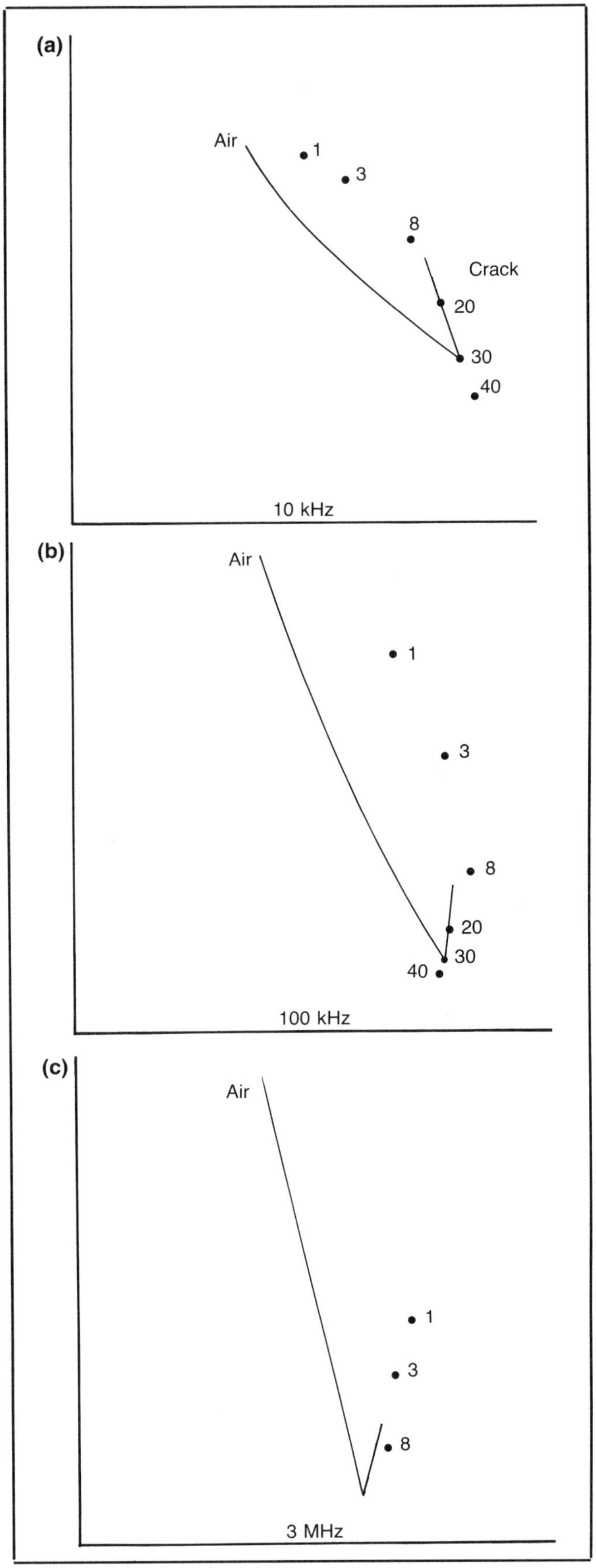

Figure 10-17. Impedance-plane phasor diagrams for deep crack in aluminum at (a) 10 kHz, (b) 100 kHz, and (c) 3 MHz. Numbers 1 through 40 indicate conductivity values in % IACS.

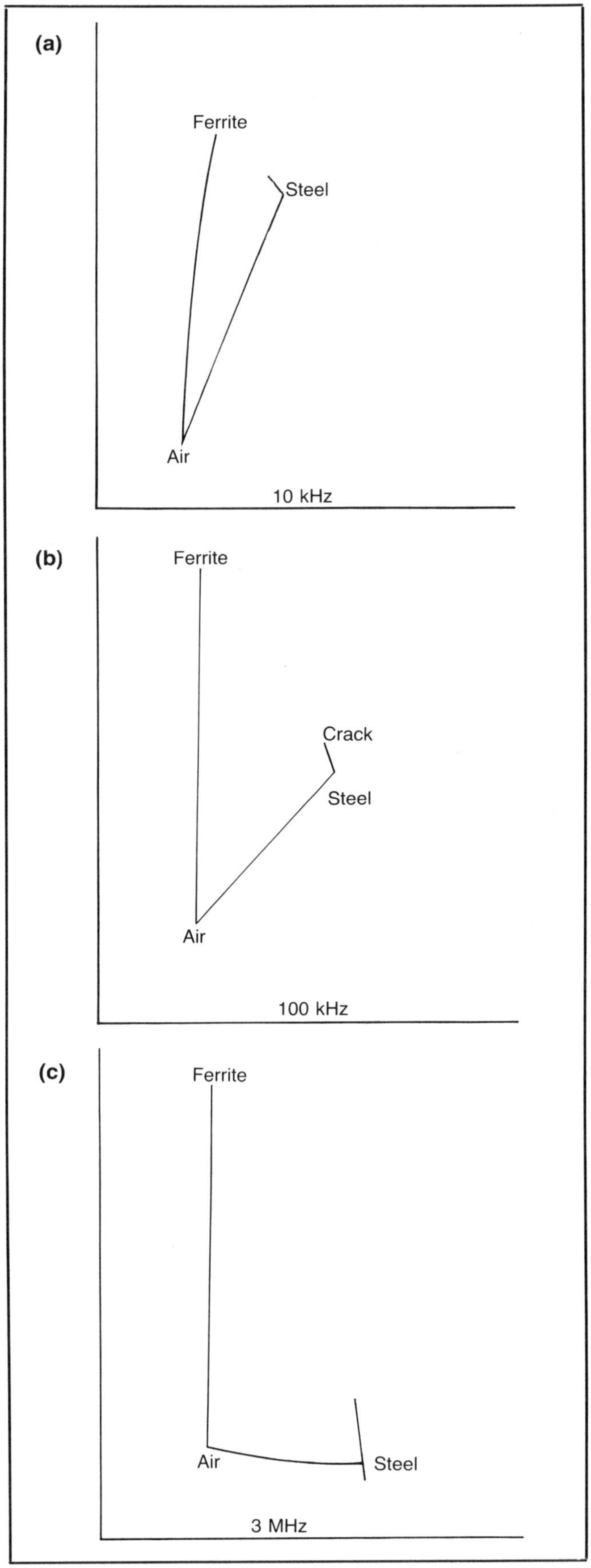

Figure 10-18. Impedance-plane phasor diagrams for deep crack in 4340 steel at (a) 10 kHz, (b) 100 kHz, and (c) 3 MHz.

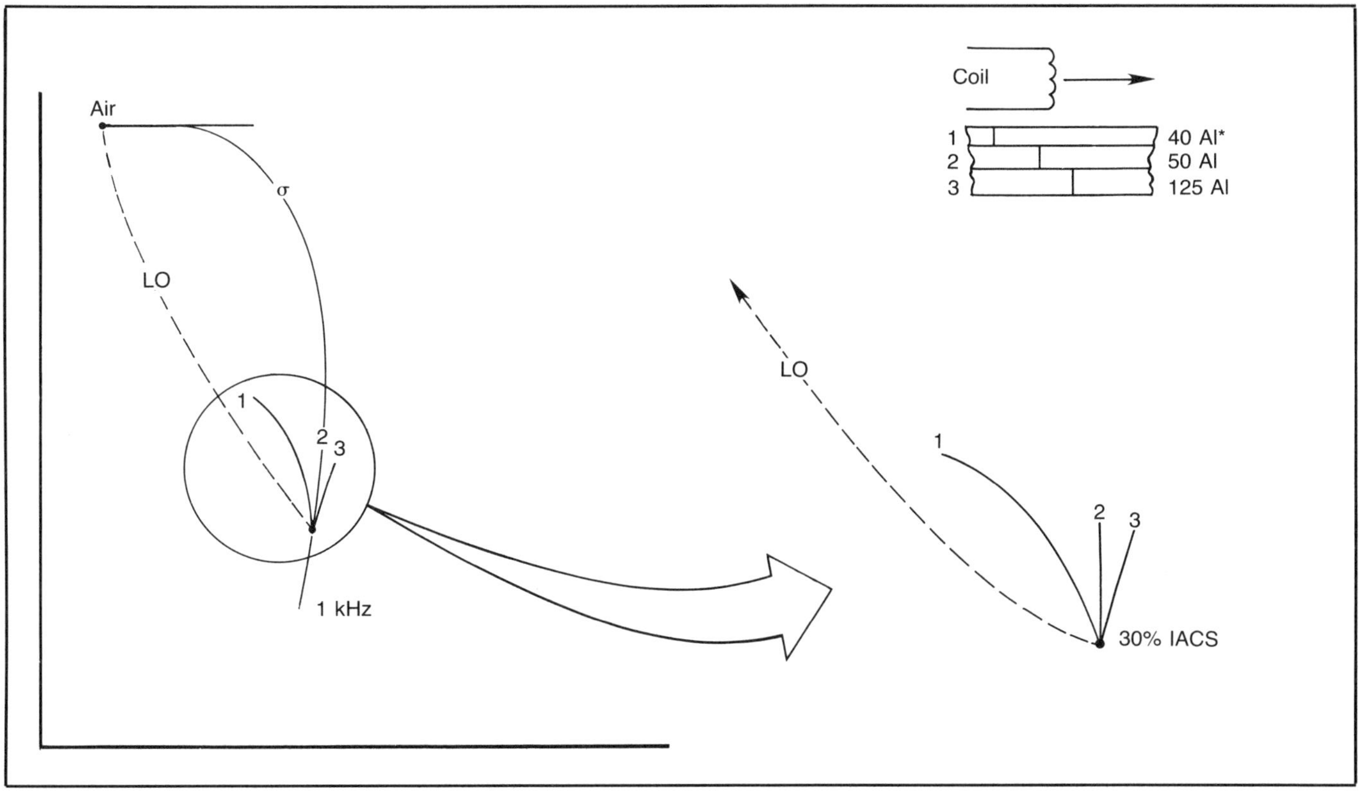

Figure 10-19. Direction of surface and subsurface cracks in aluminum on the impedance plane. Note: 2 and 3 are subsurface cracks. *Thickness in mils: 0.001 in. (0.025 mm).

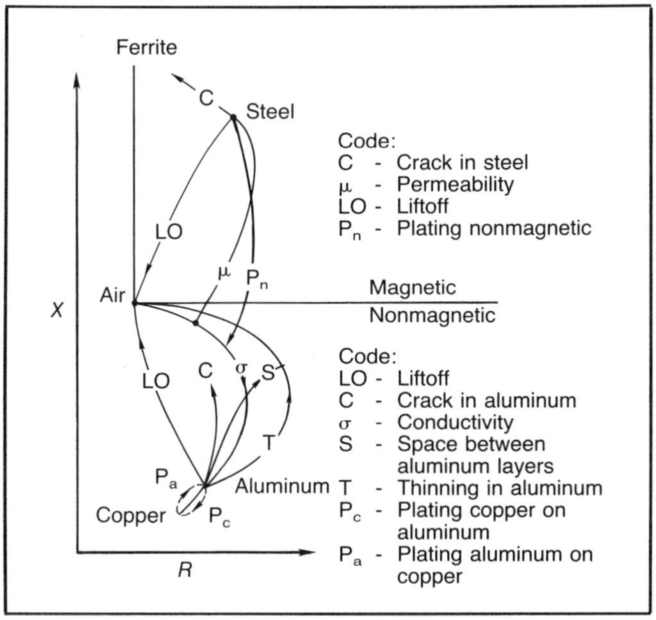

Figure 10-20. Impedance phasor changes in relation to one another on the impedance plane. Probe coil operating between 10 kHz and 100 kHz.

impedance plane and what results are to be expected, before performing any tests.

Improving Test Results

When testing with phase-analysis instruments, which present the results on a meter, a technique called "off null" balancing is used to improve the test results (discussed later in this chapter). This technique cannot be used with the CRT instruments because the flying-dot impedance point will, in many cases, be off the bounds of the CRT. Test results may be improved, however, by selecting the optimum operating frequency, by rotating the results on the impedance plane, and by increasing and/or decreasing the vertical and/or horizontal gains on the CRT. The results are rotated by turning the "phase" control on the face of the instrument (see Fig. 10-2). The horizontal and vertical gains are controlled using the dials marked "vertical" and "horizontal" (see Fig. 10-2). At 5 volts per division the gain is low (little movement on the CRT), but at 0.5 volt per division, the gain is high (considerable movement on the CRT). Two examples are shown in Fig. 10-21.

In the first example, we have the normal impedance-plane response for a crack in an aluminum sample; Fig. 10-21(a). The liftoff locus is rotated about 45 degrees to a horizontal position, as shown in Fig. 10-21(b). Because the liftoff is an unwanted response, the gain in the horizontal direction is reduced by using a setting of 2 volts per division on the CRT. To amplify the crack response in the vertical direction, the vertical gain is increased. This change in gain amplifies both the separation angle (between liftoff and crack response) and the amplitude of the crack response, as shown in Fig. 10-21(b).

The second example is for a crack in steel as normally presented on the impedance plane;

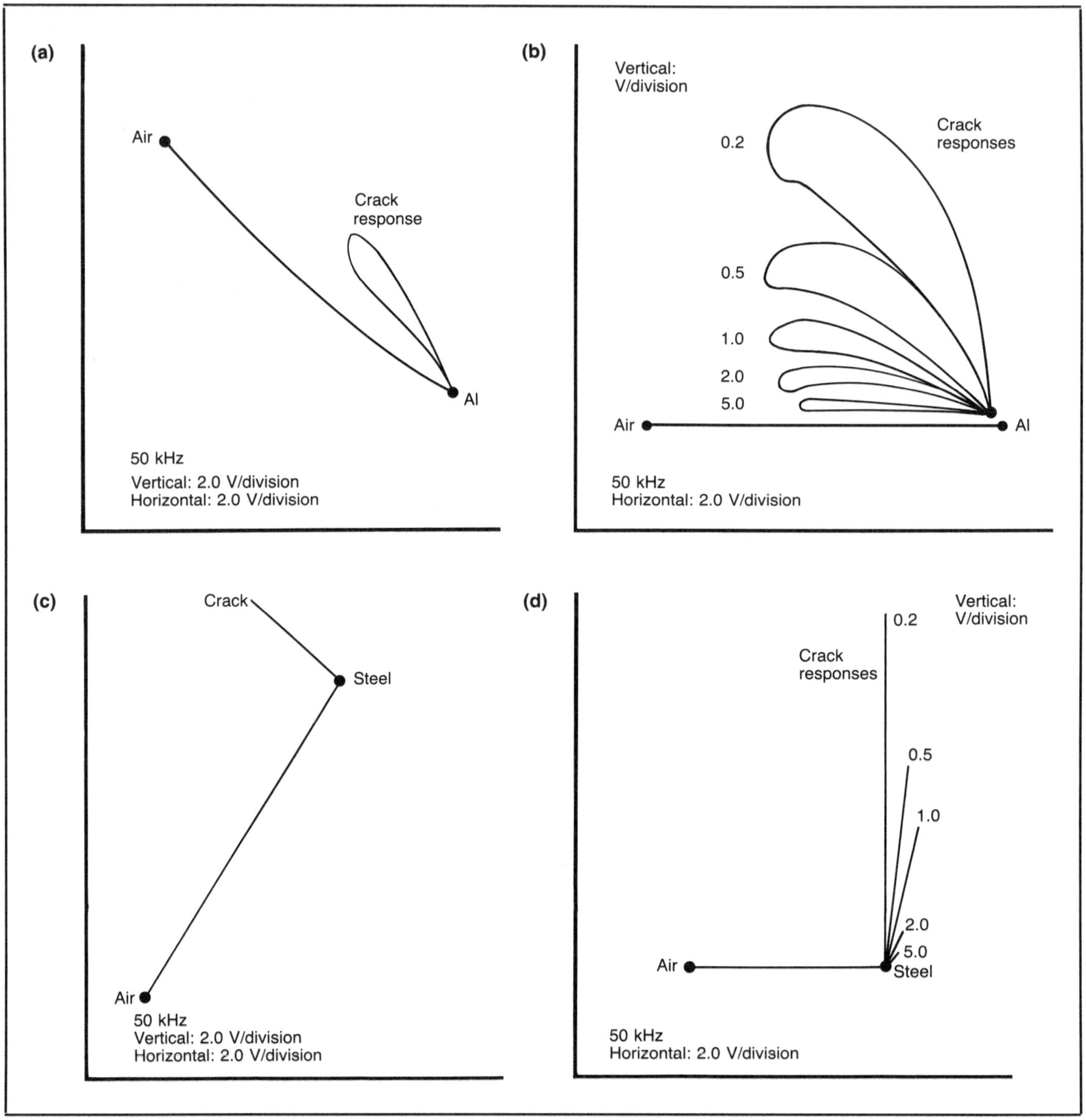

Figure 10-21. Improving test results by changing CRT vertical gain. (a) Normal impedance-plane response from crack in aluminum. (b) Crack responses at various changes in vertical gain. (c) Normal impedance-plane response from crack in steel. (d) Crack responses at various changes in vertical gain.

Fig. 10-21(c). Here again the liftoff response is rotated to a horizontal direction, with the crack response in the vertical direction. The horizontal gain is set at 2 volts per division and the vertical gain is increased to 0.2 volt per division; see Fig. 10-21(d). In this case, the phase angle does not change appreciably, but the signal amplitude is increased considerably at the 0.2-volt-per-division gain setting.

These two examples are typical methods of improving results for other test problems where the unwanted signal amplitude is reduced while the wanted signal amplitude is increased. Signals of interest may be improved or unwanted signals reduced or eliminated by employing multifrequency or mixed-frequency techniques.

Typical examples of crack responses on the impedance plane obtained using absolute probes

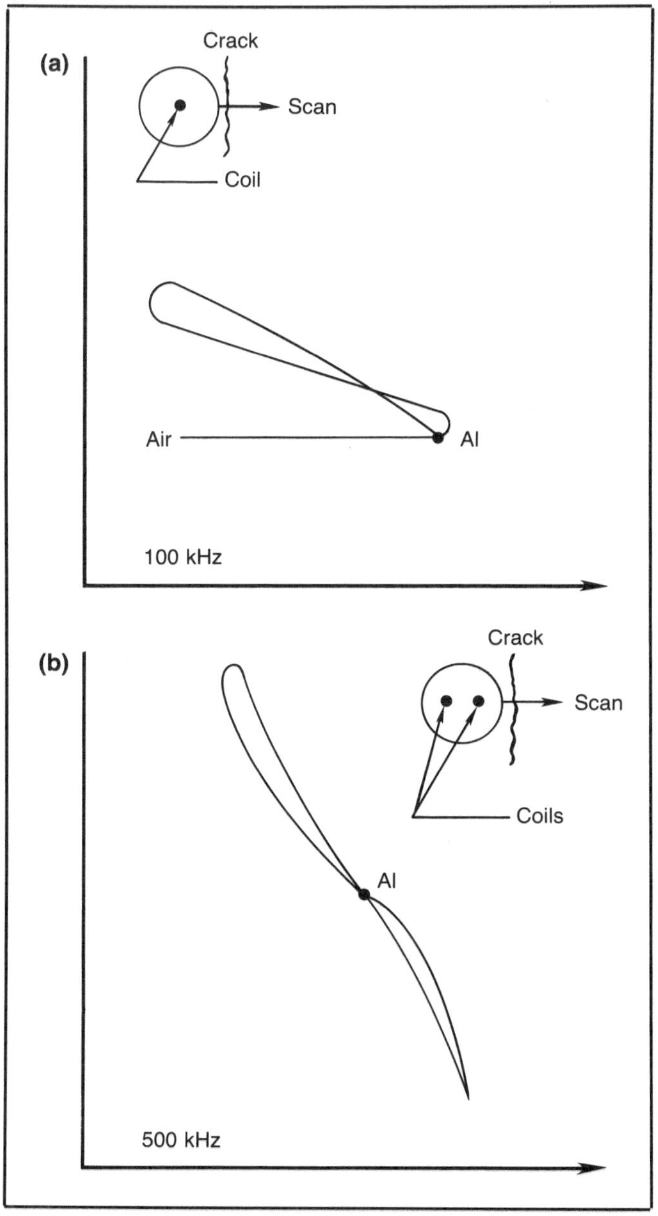

Figure 10-22. Crack responses on impedance plane for absolute and differential probes. (a) Absolute-probe crack response. (b) Differential-probe crack response.

(single coils) are given in Figs. 10-15 and 10-19. For crack detection, however, some investigators prefer to use differential probes (dual coils). Because these probes yield different responses on the CRT, Fig. 10-22 is included to show the differences. The typical single-loop crack response for an absolute bolt-hole probe inserted in an aluminum sample is shown in Fig. 10-22(a). Fig. 10-22(b) shows the typical double-loop response obtained when using a differential probe. When scanning the same cracked sample, the double-loop amplitude for the differential probe generally is double the single-loop amplitude for the absolute probe.

The choice of absolute or differential probes is generally based on operator preference. In some

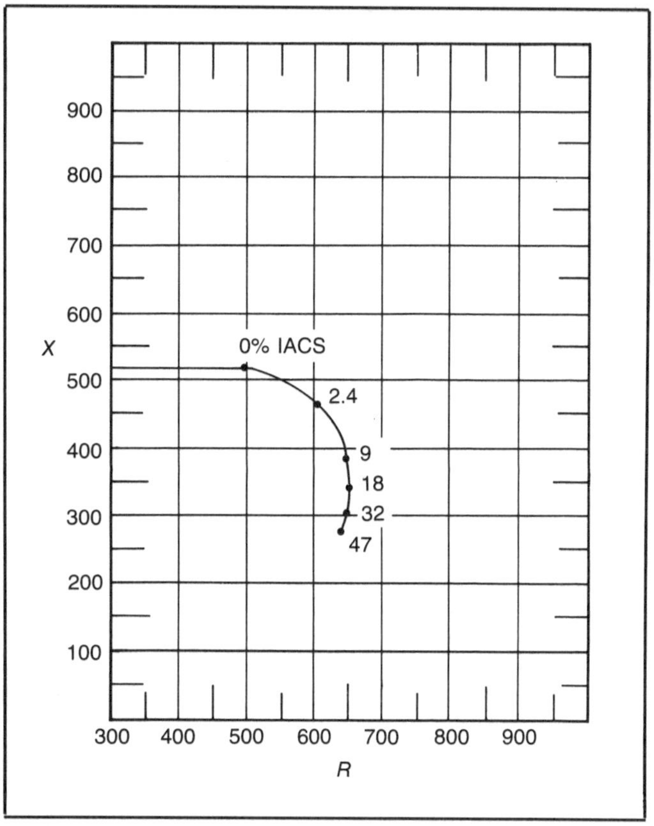

Figure 10-23. Conductivity-change impedance graph.

cases, differential probes are selected to improve the signal-to-noise ratio. Additionally, for the inspection of steam generator tubing, reports indicate that the use of absolute testing results in a substantial reduction of error in assessing the depth of flaws.

Phase-Sensitive Variable Suppression

With phase analysis or phase-sensitive metered instruments, one may balance (null) the bridge in a manner that allows the suppression of unwanted variables. This type of procedure is generally called *"off null" balancing*. These instruments contain two controls marked *X* and *R*. The *X* control balances the reactance of the circuit, and the *R* control balances the resistance of the circuit. Hence, by selecting specific values for these two variables, the bridge may be set to operate from any point on the impedance plane. Operation of the equipment in this manner allows the operator to select the operating point that permits the response to be selective in suppressing unwanted variables, while enhancing those of interest.

To accomplish the balancing operation, the operator turns the *X* and *R* controls (with the probe in air) until the balance point is reached. The meter needle is kept on scale with the balance control. At the null point, the needle will reverse direction if the *X* and *R* controls are moved past the balancing point. When the balance point is reached, the

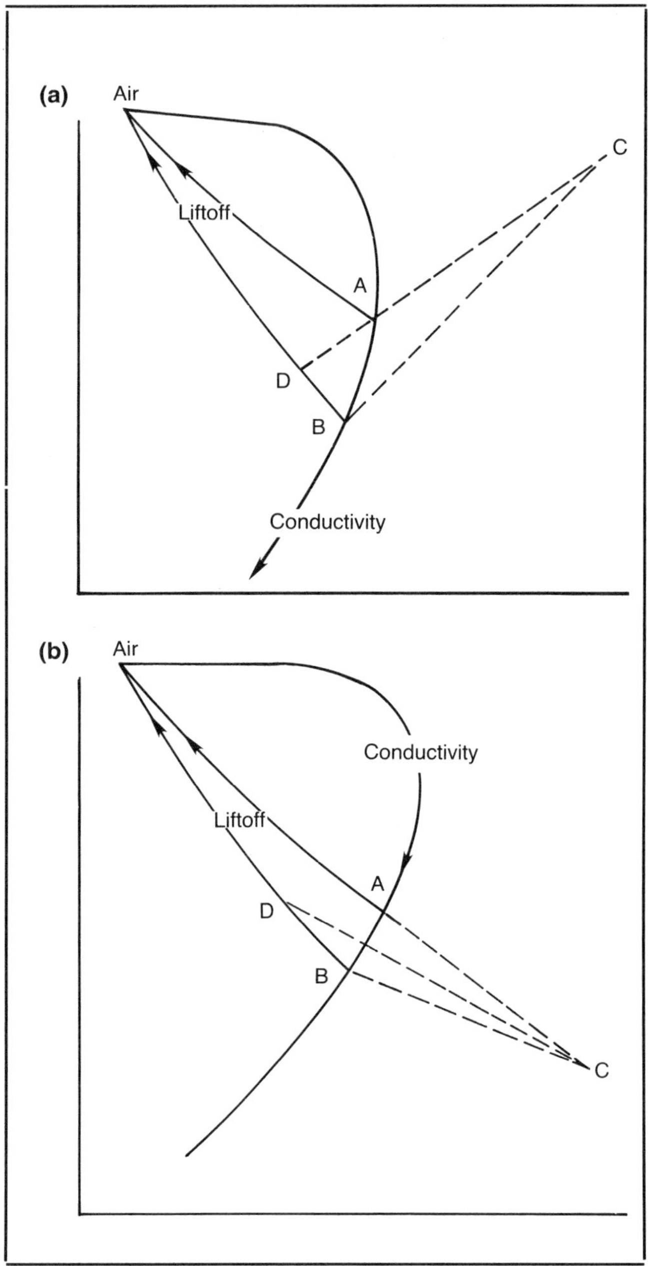

Figure 10-24. Off-null balancing on impedance graph to suppress unwanted variables while enhancing wanted variables. (a) Liftoff suppression. (b) Conductivity suppression.

operator reads the values of X and R from the dials associated with each control. For example, in Fig. 10-23, the air point balance is at $X = 525$ and $R = 500$. The probe is next placed on a 70 μΩ-cm (2.4% IACS) specimen, and the balancing procedure is repeated. In this case, $X = 460$ and $R = 610$. The procedure is repeated for any additional specimens to be tested. After all the X and R vector points have been plotted on the graph, the typical conductivity comma curve is drawn through the plotted points.

If it is desired to suppress liftoff while enhancing the separation of two alloys by differences in conductivity, the operator performs the balancing operation with the probe in air and on each of the two test specimens. Their points are plotted on the graph paper and the liftoff lines and conductivity curve are drawn on the graph.

The liftoff is suppressed by locating the instrument balancing point at a position located about 90 degrees (normal) to the liftoff line; see Fig. 10-24(a). At this point, vectors C-D and C-B are equal in length, so the liftoff is suppressed and will give very little meter movement. However, the vector C-A is shorter than vector C-B and hence point A will give low meter readings and point B will give high meter readings. The gain is then adjusted so that full-scale deflection may be obtained between the sample at A and the one at B.

Because crack responses are parallel to the conductivity loci plot (see Fig. 10-16), this same off-null point could be used to suppress liftoff while enhancing crack response on the meter movement.

If the thickness of a nonconductive coating were to be measured, one would try to suppress changes in conductivity while enhancing liftoff. In this case, the operating point (off null) is selected as shown in Fig. 10-24(b). Here, vectors C-B and C-A have equal length and changes in conductivity are suppressed. However, vector C-B is shorter than C-D, and changes in liftoff from specimen B will cause up scale readings on the meter.

All the impedance-plane variables, shown in Fig. 10-20, may be plotted on the impedance-plane graph of Fig. 10-23. The off-null point is selected normal (90 degrees) to the variable to be suppressed and parallel (or nearly parallel) to the variable to be enhanced.